D0722068

PRODUCT DESIGN METHODS and PRACTICES

PRODUCT DESIGN
METHODS
and
PRACTICES

HENRY W. STOLL

Northwestern University
Evanston, Illinois

MARCEL DEKKER, INC.   NEW YORK • BASEL

ISBN: 0-8247-7565-1

This book is printed on acid-free paper.

Headquarters
Marcel Dekker, Inc.
270 Madison Avenue, New York, NY 10016
tel: 212-696-9000; fax: 212-685-4540

Eastern Hemisphere Distribution
Marcel Dekker AG
Hutgasse 4, Postfach 812, CH-4001 Basel, Switzerland
tel: 41-61-261-8482; fax: 41-61-261-8896

World Wide Web
http://www.dekker.com

The publisher offers discounts on this book when ordered in bulk quantities. For more information, write to Special Sales/Professional Marketing at the headquarters address above.

Current printing (last digit):
10 9 8 7 6 5 4 3 2 1

PRINTED IN THE UNITED STATES OF AMERICA

Preface

This book is about designing products that are responsive to customer needs and to issues of efficient manufacture. The focus is on product design methods and practices that help ensure timely design of functionally and aesthetically appealing products that are both mechanically reliable and easy to manufacture. The book is based on material developed for use in two graduate courses on product design that I have been teaching at Northwestern University, one for engineering and management students in the McCormick School of Engineering and Applied Science and the other for manufacturing management students in a joint program with the Kellogg Graduate School of Management. It also includes material developed for use in workshops on Design for Manufacture, Quality by Design, and Creating Winning New Products that I present on a regular basis both as part of the university's executive training program and to individual companies as part of my consulting practice.

The purpose of the book is threefold: (1) to provide the reader with a comprehensive perspective of "good" design, (2) to present a road map for "doing" good design, and (3) to provide a set of structured methodologies that can be used in a systematic way to support the product design process. A fundamental tenet of the book is that, to design for quality and ease of manufacture, both the technical side and the management side of product development must be considered. Accordingly, the book integrates design principles and practices for good product design together with structured methodologies, procedures, checklists, and best management practices for implementing an efficient product realization process.

The book is intended for practicing engineers and managers who recognize that excellence in product design is crucial to the survival of manufacturing enterprises in today's highly competitive global economy and who are strongly

iii

motivated to improve the quality and profitability of their products. It is also for students who wish to improve their product design skills by developing improved understanding and insight into the complex relationship that exists between quality, cost, and productivity and how this relationship is influenced and often irreversibly fixed by product design. The book is unique in that it focuses directly on the relationship between manufacturability, quality, and design and on how to improve manufacturability and quality by design. It provides structured methodologies and practices that can be put into immediate use by product development teams. It also presents an integrated overall view of product development, design for manufacture, quality engineering, and associated best management practices and serves as a basic reference.

As a product designer, I have always believed that there are certain underlying design principles, which, if properly applied, will result in a design that is inherently robust and easy to manufacture. Similarly, I believe that there are certain design methods and practices that are particularly useful in helping the design team to do good design. One of my goals in writing this book has been to pull these ideas and approaches together into some sort of organized and useful form. They have been gathered from many sources and, in some cases, they have been enriched, adapted, and modified. I have made no attempt to create an exhaustive list of design principles or to present a comparative study of all available design methods and in this sense, the book represents my personal view of the subject. In keeping with this approach, the design methods and practices presented are not "cast in concrete" and I would hope that the reader will, in the spirit of continuous improvement, feel free to enrich, adapt, and modify them as needed to better serve specific design needs and organizational cultures.

Since becoming involved in product development and design for manufacture, I have been fortunate to have worked with and been exposed to the ideas of many people from industry and academia. Much of my original thinking was developed while I was at the Industrial Technology Institute (ITI) in Ann Arbor, Michigan, which provided an excellent learning environment in which to evolve and test new ideas and approaches to product design. While at Square D Company, I had the opportunity to try out many of the ideas discussed in this book and to see positive benefits resulting from their use. My close association with colleagues at Northwestern University and the Evanston office of IDEO Product Development has also helped to mature my understanding of the product development process. Finally, there is no doubt that my thinking has been indelibly shaped by the many excellent articles and books that have been published on this subject.

Henry W. Stoll

Contents

Contents

Total Cost Reduction

Design for Quality

PRODUCT DESIGN METHODS and PRACTICES

1

Introduction

1.1 GOOD DESIGN

This book is about "good" product design. By good design, we mean the timely design of functionally and aesthetically appealing products that have inherent high quality, low cost, and ease of manufacture. *Timely* implies the ability to quickly create products to meet new and fast changing market needs. *Functionally and aesthetically appealing* implies a design that meets well defined and clearly understood customer needs in a way that delights the customer and ensures long-term, sustainable customer satisfaction. *Inherent* implies that high quality, low cost, and ease of manufacture are ingrained properties and characteristics of the product, which has been carefully and systematically created by design.

In good product design, the term *quality* is used in its broadest sense to mean a product that provides the performance and features the customer wants, consistently conforms to established standards, is robust against hard to control variation, has long-term mechanical reliability and durability, and is easy to maintain and service. Similarly, the term *cost* refers to total life cycle cost and includes direct cost, tooling and capital cost, system cost, development cost, and the cost of time. Finally, *ease of manufacture* implies that the method of manufacture and assembly has been considered from the inception of the design and, as a result, the product and process are compatible and coordinated. This book presents a collection of design principles, strategies, and methodologies that help facilitate and insure good design.

The focus of the book is on engineered designs. Engineered designs are assembled products, equipment, devices, and hardware that have been designed to meet specific end user needs and are sold by an enterprise to customers. In this context, nuts and bolts, ball bearings, computers, electrical transformers,

1

portable phones, automobiles, machine tools, construction equipment, consumer products, and aircraft are all considered to be engineered designs.

For many firms involved in the design and manufacture of engineered designs, good design is good business; that is, it is crucial for economic success and long term survival. Good design helps keep the firm competitive and on the leading edge of fast changing markets. It helps ensure products that can be sold at prices significantly higher than manufacturing cost. And it results in products that are able to gain market share on a consistent and sustainable basis because of recognized superiority compared to best-in-class competitor products.

This book is based on the fundamental premise that there are certain underlying design principles and practices which result in good design. The goal of the book is to provide insight and familiarity with these principles and practices and to present a variety of design methods and techniques which, when applied in a structured and disciplined manner, help the design team to implement these principles and practices.

1.2 TEAM APPROACH

To insure proper consideration of all design needs and requirements, all stakeholders (i.e., interested parties) in the new product program must work together from the beginning of the design project. This means that all aspects of the design, including market analysis, process planning, facilities planning, manufacturing and automation equipment design and procurement, tooling design and procurement, and procurement of supplied parts, must be performed in a concurrent or overlapping manner as an integral part of the product design process. This concept, which involves early and constant interaction between all stakeholders in the project, has become widely known as *concurrent engineering* or *simultaneous engineering* (Fig. 1.1).

In most practical situations, concurrent engineering is best implemented using a team approach. In the *team approach*, the design is directed and coordinated by a team of individuals representing all stakeholder activities. As the design evolves, each team member stays in close communication so that changes and emerging needs are quickly relayed to each activity as they occur. This allows all activities to progress in a more parallel and overlapping manner, thereby shortening the development cycle and increasing completeness of information on which design decisions are based. Engineers and designers start work before market analysis is complete, and manufacturing and sales begin gearing up well before the design is finished. In essence, the team approach establishes the high quality communications channels needed to design the product right the first time. Effective use of the team approach is assumed throughout this book.

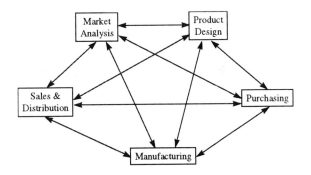

Figure 1.1 Concurrent engineering model.

1.3 STRUCTURED DESIGN METHODS

A central premise of this book is that good design can be achieved through disciplined use of structured design methods. *Structured design methods* are procedures, techniques, and tools that help guide and facilitate the solving of design problems. Design methods benefit the design problem solving process in two ways. First, design methods provide discipline and objectivity by formalizing various procedures of design. Formalized design procedures help to (1) avoid oversights and other errors that can occur with more informal or haphazard approaches, (2) broaden the scope of the problem solving activity to ensure that all viable solutions are identified and considered, and (3) document the decision making process. Second, by making the problem solving process explicit, design methods help facilitate the team approach. All members of the team can see and understand what is going on and contribute to the process. Group consensus is quickly achieved and the probability of making the best possible design decision is greatly increased. A primary goal of this book is to present a set of design methods that will help the team make the best possible design decisions at each step in the product development process.

To appreciate how design methods work, consider the simple checklist (Cross, 1994). By listing key elements or tasks, the checklist formalizes the process and makes it explicit. As items are performed, they are checked off until everything is complete. This insures that items won't be overlooked or forgotten and frees team members from the need to carry a multitude of details in their heads. Teamwork is facilitated because the checklist makes needed information available to all members of the team. Also, separate sections of the list can be allocated to different team members, saving time and improving efficiency. When the process is complete, the checklist becomes a record of accomplishment for future reference.

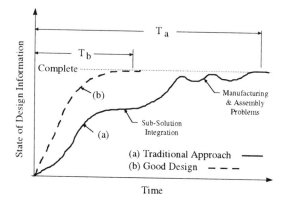

Figure 1.2 Beneficial effect of up-front planning and the team approach on the time history of a subdivided design problem.

1.3.1 Problem Solving Strategies

Problems of design can often be more effectively solved with less effort and less iteration by employing various problem solving strategies. The following strategies have proven particularly useful and are employed by most of the design methods presented in this book.

Subdivide Complex Design Problems

In many cases, complex problems of design can be solved by breaking the overall problem down into a set of defined subproblems. The subproblems are then solved in parallel and the subsolutions combined into an overall solution. This strategy has proven effective in practice because subproblems are generally more easily solved in a shorter time and with less iteration. Also, proven standard solutions for some of the subproblems may be available. In addition, by solving the subproblems in parallel, time is eliminated from the overall process and resources can be utilized more efficiently.

Planning for integration of subsolutions is often an important aspect of this strategy. For example, planning how the subsystems ("black boxes") of an electronics product are to be mounted and wired together, before the subsystems are designed, ensures that the subsystem designers have the information needed for locating electrical connectors and designing mechanical mounting details. With careful early planning and the team approach, problems and delays associated with integration, and with manufacturabiltiy and assembly problems discovered late in the project, can often be avoided (Fig. 1.2).

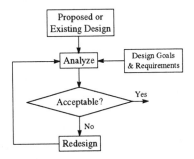

Figure 1.3 The design-analyze-redesign strategy.

Use a Structured Process

By structured process, we mean a defined step-by-step process that can be followed to solve a general class of design problems. In this strategy, the problem of design or design activity is subdivided into sequential, overlapping steps that proceed from the general to the specific. The use of structured methodologies can be very successful in practice because (1) it breaks the problem solving process down into a series of well-defined, easier to perform steps and (2) when each step is performed completely enough and well enough, major iterations between steps can be avoided. It should be noted that this strategy is also hierarchical in nature. That is, each of the steps in a structured methodology can usually be further subdivided into structured methodologies.

Design-Analyze-Redesign

The design-analyze-redesign strategy begins with an existing or a proposed design solution that is then analyzed to determine its acceptability with respect to a specified set of design criteria (Fig. 1.3). Based on the results of the analysis, the proposed design is either accepted or it is redesigned to correct deficiencies discovered by the analysis. The redesign is guided by the results of the analysis and the design criteria. The process is repeated iteratively until an acceptable design is achieved. The design-analyze-redesign strategy is very effective in practice because it leverages the inherent iterative nature of design.

1.3.2 Benefits and Caveats

Structured design methods formalize various procedures of design and make them explicit. Our goal is to present a variety of design methods that can be used at all stages and levels of design. Using these methods requires extra effort and may appear, on the surface, to be diverting time and attention away from the central task of producing a product design. So why use them? In our opinion, the answer is clear. Design methods bring structure and rationality to what can be a very open ended, random process. They force the team to think strategically about both the product that is being designed and the process being used to design it.

Use of structured design methods requires discipline and dedication from the team as well as vision and long-term support from management. In general, expect effectiveness to increase with practice and as experience is gained through repeated use. Don't become discouraged, however, if the first few attempts take longer than expected or yield less than hoped for benefits. Product design is a process that must be continually improved. Using structured design methods effectively to design better products with lower cost and less time is an important step in this improvement process.

1.4 PLAN OF THE BOOK

The chapters that follow provide a compilation of proven design methods, principles and practices that help in the creation of good product designs. Each chapter deals with a specific aspect of good product design and can be read and applied without reference to other chapters.

- Chapters 2-7 present the philosophy and considerations of good design. Chapters 5 and 6 present underlying principles of good design which provide the theoretical foundation for much of the book.
- Chapters 8-11 present a variety of conceptual design methods and approaches.
- Chapters 12-17 focus on design for efficient manufacture and include a discussion of process driven design, design for manufacture and assembly, and tolerance design. Chapter 17 presents an innovative design methodology for analyzing and developing high quality product designs that are easily manufactured.
- Chapters 18-21 consider a myriad of methods and strategies for reducing total cost. The goal of these chapters is to provide a road map for moving a product to the next plateau of cost reduction by optimizing the product and manufacturing system as a whole rather than squeezing pennies out of piece parts. Standardization strategies and strategies for increasing manufacturing robustness are particularly emphasized.
- Chapters 22-24 present a variety of methods for improving product quality by design. The final chapter in the book presents a comprehensive design review checklist to help ensure good product design.

2

The Product Design Environment

2.1 INTRODUCTION

Product design is generally conducted by a manufacturing enterprise whose primary purpose is to manufacture and sell products for a profit. Although product design is an extremely important activity performed within the manufacturing enterprise, it is just one of a variety of activities which must be performed well if the firm is to survive and prosper in the long term. How these other activities and concerns enable and/or constrain the product design process is therefore of key importance in good design. In this chapter, we survey the multitude of business and organizational considerations that influence and help define good product design.

2.2 STANDARDS

The starting point for good design is a company-wide attitude that values and supports the development of high quality new products. Often, this support is provided by an overall directive that motivates all employees of the firm and puts in place a supportive organizational framework that values product design. This overall directive is most effective when formulated as a set of "expectation standards" which convincingly communicate basic beliefs such as "excellence", "to be the best", or simply "quality" (Bloch and Geitner, 1990). "ISO 9000 Certification" and "6 σ Quality" are two widely publicized examples that illustrate how a firm can formulate high expectation standards within its organization. Expectation standards must generate in the firm the will to pursue good design. Once these expectation standards exist, it becomes possible to allocate the time and resources needed and to develop the design process, design methods and organizational support necessary for good design.

2.3 MARKETING AND MANUFACTURING STRATEGY

How a product is designed depends in large measure on the way the product is to be manufactured and brought to market. Companies that understand these issues often develop strategies for designing and marketing their products. In companies that seek competitive advantage through product design, a wisely conceived business strategy can be of tremendous value in providing guidance and direction to the product design process. The challenge of good design is to identify the best choices out of all the choices that are possible. The value of a clearly articulated strategy is that it helps to eliminate and narrow choices, makes the best choices more obvious, and constrains choices that would otherwise be arbitrary.

To appreciate how business strategy drives design, consider two different personal computer manufactures. The marketing strategy of company "A" is to sell several different product models, differentiated on the basis of cost and performance, through large retail outlet chains and other mass retailers. To support this marketing strategy, a manufacturing strategy is developed in which all models are mass produced on one assembly line and delivered in scheduled shipments to retail outlets using an optimized distribution system. With the marketing and manufacturing strategies thus determined, the design strategy becomes obvious: design the product so that all models can be mass produced interchangeably on one line and packaged such that specified quantities of each model can be easily sorted and palletized for scheduled shipment to specific retailers.

The marketing strategy of company "B", on the other hand, is to sell its product directly to the end customer and in this way allow the customer to customize his or her computer via a special order. This marketing strategy needs to be supported by a manufacturing strategy which facilitates easy manufacture and delivery of customized product together with a design strategy which makes the product easy to customize at remote distribution sites. Because of the different strategies used, design choices and decisions for each company are quite different. What constitutes a good design for company "A" is clearly different from that for company "B".

Design strategy depends on the marketing strategy and the manufacturing strategy (Fig. 2.1). As a result, design strategies differ from firm to firm and product to product. Design strategies help guide design decision making. Therefore, a good design decision in one situation may be a poor one in another. Similarly, the principles and practices of good design discussed in this book may be more applicable to some situations than to others, depending on the design strategy involved.

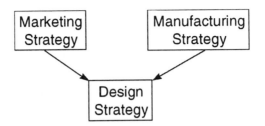

Figure 2.1 Design strategy depends on the firm's marketing and manufacturing strategies.

2.4 PRODUCT DESIGN SITUATIONS

Design projects can generally be categorized depending on the newness of the product and on how free the design team is to select new manufacturing methods and processes (Fig. 2.2). A *new product* is one that involves significant change from existing models or versions of the product. Styling, working principle, materials, and technology can be changed or be different from past practice. An *existing product*, on the other hand, is a current product that is redesigned to reduce manufacturing cost and/or to correct and improve product performance and quality. New models or variants of existing products may also be designed to meet niche or new customer or market needs.

Similarly, a new product design can be manufactured using existing methods and processes or new methods and processes can be used. A *new manufacturing method or process* is one that has not been used previously by the company to produce the product being designed. An *existing manufacturing method*, on the other hand, is one that is currently being used in production. Designing a product which is to be made using new manufacturing methods and processes is quite a different problem from the design of a product which is to be produced in an existing, well understood, and well characterized production facility.

With these concepts in mind, we can define four design project situations based on the possible combinations defined by the matrix in Fig. 2.2. These situations can be thought of as forming the corners of a field in which most product design projects are located.

1.   A new product and a new manufacturing method.
2.   A new product and an existing manufacturing method.
3.   An existing product and a new manufacturing method.
4.   An existing product and an existing manufacturing method.

		Method of Manufacture	
		New	Existing
Product	New Product Design	Situation #1	Situation #2
	Existing Product Redesign	Situation #3	Situation #4

Figure 2.2 Product and manufacturing constraint matrix.

Most product design projects fall somewhere within these extremes. Each new design project offers different opportunities and challenges for good product design. Examination of these extreme situations provides insight into the variety of design approaches and potential benefits that are possible.

2.4.1 Situation #1

This design project situation is the best possible situation for creating a good product design because neither the product nor manufacturing method are constrained by past practices or fixed investments. Unfortunately, because both the product and the process are new, product knowledge is probably the least complete in this particular type of design situation. Also, time and resources may be short. The challenge is therefore to take the time, up front, to correctly understand the customer requirements and demands, identify and select the best product concept from both the end customer's point of view and the company's point of view, and then design the product and method of manufacture as a coordinated system.

2.4.2 Situation #2

This situation commonly occurs in vertically integrated companies that produce a range of products using well-established manufacturing methods. The great opportunity offered by this situation is to design new products which readily map into the existing production facility with the least disruption and cost. The challenge in doing this is to correctly understand and characterize the manufacturing constraints that exist and then to package this information in a form that is readily usable by the design team. One possibility is to develop a *designer tool kit* which provides design rules and guidelines, physical examples and models, computerized design aids, and other specific information about the manufacturing facility in an easily used form. Such an approach is an example of process-driven design. *Process-driven design* can be defined as a product design strategy in which the method of manufacture is specified "up front" as a design requirement.

2.4.3 Situation #3

This situation is common in companies that have existing product lines that meet well-defined market needs. It is not unusual for a manufacturing company which has been in business over a long period of time to manufacture and sell literally hundreds to hundreds of thousands of different products. We refer to these as *mature products*. Such products typically (1) have been designed and developed one at a time and tooled many years ago, (2) are part of a portfolio of products which has been expanded through evolutionary and chronological developments, and (3) do not embrace much compatibility, standardization, or modularity.

A significant opportunity in this design situation is to leverage the concepts of standardization and modular design together with flexible automation and advanced manufacturing technology to achieve the following goals:

- Family of products derived from one "maximum" design.
- All models on one production line.
- Resetting time for type change in minutes.
- New variants quickly and easily designed and introduced.
- New technology, materials, and processes easily incorporated.

Achieving these goals usually requires redesign of the existing product. The objective is to create a coordinated and unified product family that makes all subsequent product designs (within a given product family) situation #2 projects. The challenge is to redesign the existing products in such a way that these goals can be achieved while, at the same time, adhering to defined parameters regarding functionality, working principle, outward appearance, and so forth. This is especially true for industrial products (e.g., circuit breakers, valves, and switches) which are often components in larger systems and are also often subject to code and other regulatory approvals.

In some mature products, it may be possible to replace costly or failure prone components with new technology, materials, or processes. For example, a complex weldment involving several parts could be replaced with a thin wall, light-metal sand casting. The development of rapid prototyping technologies, such as stereo lithography and layered object manufacturing, has made casting a viable alternative, even in applications involving low production quantities.

The identification and re-engineering of conversion parts is an important aspect of situation #3 for reducing cost and improving quality, especially in products where a total redesign is not economically feasible.

2.4.4 Situation #4

This situation most commonly occurs when problems develop after a new product has been introduced into manufacture or when initial manufacturing costs are higher than expected. Resolving this situation can be particularly difficult because of the constraints imposed by the existing product design, tooling and equipment investments, and manufacturing method. Often, only minor design changes can be considered and resources are limited. And, any change that is made can necessitate other changes that sooner or later result in suboptimal design and undesirable compromise.

Finding an effective design solution in this situation requires that the firm have an attitude of *continuous improvement*. Such an attitude recognizes that, even after the product is in production, improvements should be made on an ongoing basis to maximize the goodness of the design. The willingness of a firm to continuously improve is often engendered by the firm's expectation standards discussed previously. Continuous improvement is implemented by (1) simplifying the product and process to reduce cost, cycle time, and waste; (2) identifying features and product improvements that will further increase customer satisfaction and/or correct unforeseen problems; and (3) improving the design process so that better products can be introduced faster.

Continuous improvement essentially acknowledges that the benefits of global optimization of product and process (lower cost, better quality, improved production, and design flexibility) always justify, in the long term, the incremental cost required to achieve it. It also implies that incremental improvements are usually most effective. Introducing frequent small improvements based on customer reactions or improved manufacturability is less risky than undertaking a major redesign of the product to correct unforeseen problems.

2.5 THE MANUFACTURING ENTERPRISE

Products are designed, manufactured and sold by a business or manufacturing enterprise. To appreciate the relationship between design, manufacturing, and business, consider the basic actions that must take place to operate a manufacturing firm. Four groupings of "main line" work are generally necessary: decide what customers want, set up the factory to make it, produce it, and sell and service it. These work groupings can be combined to form a simple model of the manufacturing enterprise (Fig. 2.3). Tracing the "main-line" generic workflow that connects these activities together yields two major

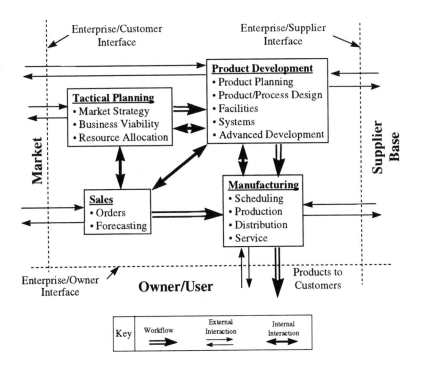

Figure 2.3 Simplified model of the manufacturing enterprise (Chapter 7, Ettlie and Stoll, 1990).

cycles on which most design and manufacturing practices focus: (1) the order-to-delivery cycle which involves the receipt of orders and production of existing products, and (2) the new product introduction cycle which involves planning and introducing new product changes into production.

The order-to-delivery cycle is the operations side of the business; it generates the day-to-day cash flow for the firm by producing and distributing products, assuring product quality, and providing the interface between the company and its customers in the form of sales and service. Major activities typically include order fulfillment, production, and service. The order fulfillment activity brings in product orders (sales) and forecasts production requirements. Production converts sales orders into products. It includes all of the personnel, operations, processes, tooling, and equipment involved in the day-to-day manufacture of existing products including procurement, raw material and supplied component receiving and inspection, part manufacture, assembly, testing, and distribution of manufactured products. Service provides product maintenance, repair, and disposal services and support to the customer.

The new product introduction cycle involves two major activities: tactical planning and product development. Tactical planning determines the "what" and "when" for product design and development and quantifies the needs that guide design and manufacturing. Product development translates the "what" into "how" by defining the new product designs to be introduced into production and providing the physical hardware, systems, information, training, and guidance to make new processes, tooling, and equipment work in production. Tactical planning is typically performed by business planning and marketing groups while product development is typically performed by product planning, technology development, product engineering and manufacturing engineering groups.

The focus of this book is on the product development activity depicted in Fig. 2.3. It is important to understand, however, that product development is carried out in the context of the larger manufacturing enterprise and that the other activities of the manufacturing enterprise impose a variety of boundary conditions and constraints on the product development process and on the product designs that are developed. Often, the ability of a firm to perform product design effectively and competitively depends on how the interactions that occur between the product development activity and the rest of the enterprise are facilitated and managed.

2.5.1 External Interaction

External interactions include the firm's interaction with its customers and with its supplier base. These interactions can significantly effect the ability of the firm to design and market successful products. External customers include those who purchase, use, regulate, distribute, install, service, and ultimately dispose of and recycle the company's products. Interaction with external customers provides the understanding needed to correctly design and market products that delight and satisfy customers. The supplier base consists of those organizations from which the firm purchases raw materials, components, subassemblies, and other goods and services required in the manufacture of its products. A good working relationship with the supplier base helps both to control cost and to insure the availability and quality of supplied materials and components.

2.5.2 Design-Marketing Interaction

The *design-marketing interaction* occurs between the product development, tactical planning, and sales activities (Fig. 2.3). The output of the tactical planning activity is an initial product specification (or business plan). The product specification typically provides business justification for the new product design program and sets design goals in the form of product requirements (that is, detailed specifications for features, performance, compatibility, ergonomics, and appearance or styling). It also specifies the target manufacturing cost and margin, unit volume, expected capacity utilization, and timing of introduction. The interaction occurs as the marketing and product design activity work together to accomplish the design task set forth by the product specification. A well-managed design-marketing interaction is critical in good product design. This is because the requirements definition and early conceptual design of a new product or redesign of existing products often demands a close working relationship between the marketing and product design activities.

2.5.3 Design-Manufacture Interaction

The *design-manufacture* interaction occurs between the product design and manufacturing activities (Fig. 2.3). Because the design of the product and its method of manufacture are intimately connected and interdependent, this interaction is extremely complex. It cuts across all aspects of the manufacturing enterprise, and occurs at the piece-part, assembly, systems, business, and organizational levels of the firm. Piece-parts and assemblies must be compatible with the processes, equipment, and tooling used in their manufacture. At the systems level, the product design and manufacturing system must be coordinated and optimized to reduce total cost and maximize quality. At the business level, the new products introduced into production by product development must map into the production environment of the company with the least amount of disruption and expense. At the organizational level, design engineering and manufacturing engineering must work together in a coordinated and cooperative manner to insure optimization of the manufacturing system as a whole.

The design-manufacture interaction is easiest to see at the piece part level and assembly level, where ease of manufacture depends on how the product and components are designed. The interaction is more difficult to comprehend at the systems level, and yet it is at this level that these interactions have the most long-term impact and far reaching consequences. Consider, for example, the sheet metal used in making an automobile. From the point of view of design, a wide selection of different sheet metal thickness, roll widths, coatings, and material properties is desirable so that the most optimal design can be achieved with respect to structural rigidity, weight, and cost. From the manufacturing point of view, however, the ability to process a wide selection of different thickness and types of sheet metal means added cost and complexity in purchasing and supplier relations, material handling and storage, processing equipment and tooling, setups and operations, and so forth. What is needed is the right balance which gives design the flexibility it needs while still allowing manufacturing the ability to simplify and standardize its operations. How a firm manages systems level interactions such as this can greatly effect its long term economic viability and competitiveness.

The design-manufacture interaction also includes a variety of business considerations, often referred to as "make or buy" decisions. One approach a firm can follow is to assemble its products using parts and components purchased from suppliers and vendors. The firm may pay more for individual parts and components, but it avoids the direct and long-term indirect cost of making the parts internally and gains the advantage of being able to select materials and processes which best meet the functional needs of the new product. It should be noted that this approach depends on effective management of supplier relations. For companies that utilize this strategy, supply-chain effectiveness has joined product quality and time-to-market as a key competitive differentiator.

An alternative approach is for the firm to make many of the parts used in its products. In vertically integrated companies, new products may need to be designed to be compatible with available "in-house" tooling and equipment. This can impose constraints on material and manufacturing process selection. For example, the cost and difficulty of introducing a spot welded structure may be overwhelming for a company that has historically manufactured using the arc welding process. Similarly, switching to an all plastic housing may prove difficult for a company which has traditionally made sheet metal housings and, as a consequence, has invested heavily in sheet metal forming equipment and expertise.

Most established companies fall somewhere in between these extremes, sourcing some parts and making others, depending on the economics involved. In most, if not all of these firms, the design of new products is therefore constrained in one way or another by existing resources and fixed capital investments in tooling and equipment. In other words, in many practical situations, in-place facilities can be barriers to new product innovation.

2.6 MANAGING INTERACTIONS

The key to making high quality design decisions is to have needed information and input readily available when the decision is made. This requires exceptionally clear and high quality communications channels between all activities involved in new product development. Such channels are created when the company recognizes that design is not the responsibility of design engineering alone. Design of a new product is a company-wide initiative requiring input from many disciplines. Marketing needs to be actively involved throughout the project to provide access to external customers and assist with the assessment of customer reaction to design ideas. Similarly, manufacturing must be involved from the start to ensure that manufacturing constraints and requirements are properly considered. A variety of other activities must also be involved. For example, purchasing needs to ensure that timely supplier input is obtained and that sourcing options are properly considered.

There are many barriers to effective interaction between various functional organizations within and outside the firm. Most of these barriers fall in one of the following categories:

- **Inequality of disciplines.** In some companies, product engineering is king, while in others marketing or manufacturing or some other discipline controls. It is important to recognize inequalities that may exist and to balance these inequalities with processes and structured methods that help ensure and promote good communication.
- **Political problems for management.** Who is the boss? Who has final say? How is team behavior to be rewarded? Who controls the budget? Who sets the schedule? How should suppliers be selected and involved in the early stages of the project? Questions such as these must be clearly and unambiguously answered if good organizational communication during product design is to be achieved.
- **Evolutionary products which are hard to change.** Such products tend to engender functional fixedness and organizational "silos" that isolate functional groups and block effective communication. Often, structured methods such as "design for manufacture and assembly" can help break down these walls.
- **Cultural attitudes.** Effective communication depends on openness and a willingness to listen. In some situations, underlying cultural attitudes can become a serious obstacle. The following examples typify attitudes that signal cultural problems:

    - "We've always made it this way"
    - "Give me my targets and let me do my thing"
    - "You don't understand the problem"
    - "We don't have time for this"
    - "We're the market leader, we must be doing something right"

Efficient interaction between functions can be facilitated and nurtured in a variety of ways. Some of the most effective of these include:

- Management commitment to methods and practices that foster good communication and interaction, with continued encouragement and reinforcement.

- Close physical proximity of design, manufacturing, marketing, purchasing, and other key disciplines within the organization.
- A company-wide focus on the excellence of the product, not on who is responsible for what.
- Leaders who exhibit cooperation and model the proper working environment.

2.7 KEY TAKEAWAYS

There are many underlying organizational issues that surround product development. How these issues are resolved can often make the difference between good design and mediocre (or unacceptable) design. There are several actions that the firm can take to help smooth the way and facilitate good product design.

- Good product design is facilitated and encouraged by a company-wide standard of excellence. Such a standard helps provide the time and resources needed and creates a corporate culture that expects and nurtures good product design.
- Good product design is facilitated by a clearly articulated design strategy. A properly formulated design strategy should be based on the firm's marketing and manufacturing strategies and should implement its long-term business goals. A properly formulated design strategy provides design guidance and direction, resolves trade-off concerns, and defines a logical product architecture for maximizing profitability of the manufacturing enterprise.
- Each design project, both large and small, is constrained by the design situation in which it is performed. By understanding and appreciating the particular set of business, organizational, and cultural circumstances that surround the project, the team can maximize and fully leverage the product improvement opportunities that are available.
- Successful product development depends on effective and timely communication with all entities and activities, both internal and external, that provide input to and/or constrain the product design. This means that the team must interact efficiently with external customers and suppliers and with internal activities such as marketing, research and development, sales, purchasing, manufacturing, and service. Effective communication requires high quality communications channels. These channels should be established at the start of the project and should remain in place throughout its duration.

3

Design for Profitability

1.1 INTRODUCTION

The ultimate goal of a company that designs, manufactures, and sells products is to maximize profit. We can express this mathematically as,

$$Profit = Sales\ Revenue - Total\ Cost - Taxes \rightarrow Maximum \qquad (3.1)$$

where

$$Sales\ Revenue = V\ (Selling\ Price) \qquad (3.2)$$

$$Total\ Cost = V\ (m + l + p) + F + S + D + T + B \qquad (3.3)$$

and

V = lifetime product volume
m = unit material cost
l = unit labor cost
p = unit cost for production resource usage
F = fixed tooling and equipment cost
S = systems cost
D = development cost
T = time cost
B = business cost (e.g., administration, advertising, etc.)

From these relationships, it is obvious what the firm must do to improve profitability: (1) increase demand for the products it sells and (2) reduce total

19

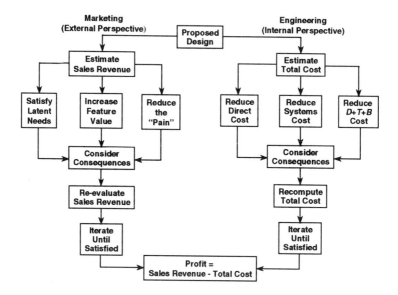

Figure 3.1 Design for profitability seeks to simultaneously maximize sales revenue and minimize total cost.

cost associated with designing, manufacturing, selling, supporting, and recycling them (Fig. 3.1). Increased demand means increased sales revenue because more units can be sold for a higher selling price. Reduced total cost means more sales revenue, including that generated from increased demand, can be converted into profits. A major focus of this book is to explore how profitability can be influenced and improved by making well-informed early product design decisions.

Product demand, selling price, and manufacturing cost are determined in large part by the product design. To be in demand and to command a high selling price, the product must meet a well-defined and understood market need in a way that delights and satisfies the customer over the long term. Good design influences demand and selling price by satisfying latent needs, increasing feature value, and reducing the "pain" associated with use of the product. *Latent needs* are explicit or implicit customer requirements that are neither fulfilled nor commonly articulated and understood. Satisfying latent needs increases demand and causes the product to sell for a higher price because the product satisfies needs that have not been previously anticipated. Similarly, correctly understanding what product features are most desirable and including these in the design causes the selling price to increase because the customer is getting more of what he or she wants and values. Is a forth door on a mini-van desirable and if so, how much could the selling price of the mini-van be increased if it has this feature? Correctly answering questions such as this is a key aspect of product pricing. Finally, assuming competing products all meet an intended need, the product that has the least "pain" associated with owning and using it will command the highest selling price. Zero defects, high reliability, ease of use, efficient operation, and so forth make the product more desirable to purchase and own.

How the product is divided into assemblies of designed and standard components establishes much of the total cost associated with the product's manufacture and life cycle support. It also defines the quality risks that must be dealt with. Most importantly, it creates many of the conditions that either constrain or facilitate the company's ability to grow demand and reduce total cost over the product life cycle. Design decisions regarding part decomposition affect not only direct cost, but indirect cost as well (Fig. 3.1). One of the challenges in designing for profitability is the limited ability available in most companies to estimate indirect cost associated with a particular product or to estimate the impact of early design decisions on indirect cost. Hence, developing insight into how design decisions effect total cost is another overarching consideration that pervades many aspects of this book.

3.2 TOTAL PRODUCT VALUE

Those who have designed products to meet specified needs know that, in most cases, many different designs will work. The challenge of design for profitability is to identify, out of all the choices possible, the best design from a functional, quality, cost, manufacturability, and business point of view and to do this as efficiently as possible. Design for profitability therefore requires a very broad and holistic perspective. Most importantly, quality, cost, and time must be understood and treated in an integrated and coordinated way.

To facilitate and encourage an all-encompassing, holistic type of thinking, it is convenient to think of design for profitability as a maximization of total product value, that is

$$Total\ Product\ Value = \frac{Total\ Product\ Quality}{Total\ Cost\ x\ Total\ Time} \rightarrow Maximum \qquad (3.4)$$

Each of the terms in Eq. (3.4) is defined as follows:

Total Product Quality: the totality of features and characteristics of a product including the product's design, manufacture, distribution, sale, service, use, and disposal that bear on its ability to satisfy stated or implied needs.

Total Cost: the sum of all costs, both direct and indirect, that result from the design, manufacture, distribution, sale, service, use, and disposal of the product over its life cycle.

Total Time: the composite of all times that are effected or determined by design decisions, including design and development time, manufacturing lead time, operation time, useful life, service, maintenance, and other support time.

In essence, Eq. (3.4) states that total value of a product is improved as the total product quality is increased and the total cost and total time decreased. It also implies that design for profitability is, in fact, a global optimization of the manufacturing system as a whole since the term *total* includes all aspects of the product's design, manufacture, and life cycle use. The task of finding practical design solutions that maximize total value and are still acceptable from all standpoints is not an easy one. This is because quality, cost, and time are coupled in complex ways, which makes it difficult to treat each term independently. Serviceability, for example, is a dimension of total product quality that also involves cost and time. Note also that a product that has a very low total cost but a high total time, or high total cost and low total time, may be less desirable than one which better balances cost and time. Finally, the practicality of a particular design solution is highly dependent on business conditions, organizational constraints, and available resources. Each term in Eq. (3.4) depends greatly on the design decisions that are made during the creation of a product. The ramifications of each term are discussed in detail in the following sections.

3.3 TOTAL PRODUCT QUALITY

Each design decision, both large and small, contributes in some way to product quality. To design for total product quality, it is necessary to decompose the concept of total product quality in a way that allows the design team to systematically focus on quality as a design objective. There appears to be many ways of doing this. For example, Garvin (1987) defines eight dimensions of product quality: performance, features, reliability, conformance, durability, serviceability, aesthetics, and perceived quality. Although this decomposition gives the design team a lot to think about, it doesn't really help the team focus on quality in a systematic way.

To facilitate a systematic approach, perhaps a more useful decomposition is to view total product quality in terms of what is important for good design. In good design, we are concerned with (1) how well a product design satisfies customer requirements, (2) how well the product performs over time and during

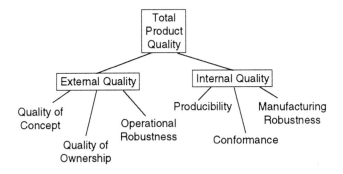

Figure 3.2 Components of total product quality.

use, and (3) how easy the product is to design, manufacture, distribute, sell, service, and support in the field. In other words, what will cause the customer to select or purchase the product? What will delight and satisfy the customer as an owner and user of the product? And, what will make it worthwhile for the firm to design and sell the product?

These concerns can be used to decompose total product quality into "external" and "internal" components (Fig. 3.2) to provide two essentially independent views of product quality. Both the external and internal qualities of a product are strongly influenced by design decisions, but in decidedly different ways. *External quality* is the component of total product quality that effects sales demand and selling price of the product. It typically depends on the design concept, the working principle, the feature level, and the styling of the product. *Internal quality*, on the other hand, is the component of total quality that effects total cost. As we shall see in subsequent chapters, more than anything else, internal quality depends on the way the product is decomposed into standard and designed subassemblies and components.

Distinguishing between external and internal quality helps focus the design team on those aspects of total product quality that can be strongly influenced by design. Each component is largely determined by the details of the product's design as well as by the design process utilized in creating and deciding these details. Also, many of the factors that characterize these qualities can be inferred in a relatively straightforward manner as part of the design process.

We employ a twofold strategy for using these two qualities to achieve high total product quality by design. First, we determine what factors or design considerations effect or contribute to each quality. We then seek to understand how these factors can be adjusted or influenced by design to maximize total product quality. This approach is developed for each quality in the following sections.

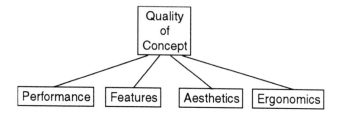

Figure 3.3 Quality of concept is a measure of how well the product delights and satisfies the customer. One way to maximize this quality is to make each factor better than best in class competitor products.

3.3.1 External Quality

External quality is a reflection of how well the product satisfies customer requirements. As discussed above, the term "external" is used because this quality is of primary importance to the customer who purchases and uses the product. As shown in Fig. 3.2, external quality is composed of three components: quality of concept, quality of ownership, and operational robustness. Each of these is discussed as follows.

Quality of Concept

Quality of concept is what makes the product desirable to the end customer and causes him or her to select it. For most products, this quality is a composite of performance, features, aesthetics, and ergonomics (Fig. 3.3). Performance is the product's primary operating characteristics. Features supplement the basic product function. Aesthetics and ergonomics involve how the product looks, feels, sounds, and interacts with the user and are often determined by the styling concept or industrial design of the product.

What constitutes an acceptable quality of concept is difficult to characterize since what is acceptable depends on the nature and type of product, the level of customer expectation, and the vision and philosophy of the firm. For example, a luxury car is judged by a different standard from that of an economy car. Also involved are elusive concepts such as perceived quality which is tied to the company's reputation and involves an unstated assumption on the part of the customer that the new product will have the same quality as an established product. Ultimately, acceptable quality of concept is determined by sales success in the market place. If a product has an acceptable quality of concept, it will generally sell well.

To maximize the quality of concept, we seek to design the product so that it is superior to the best in class competitor products with respect to each factor.

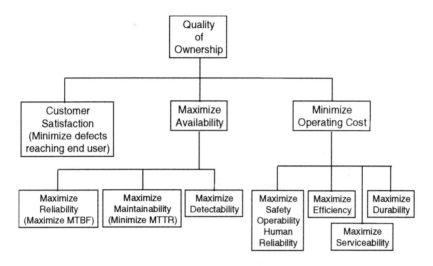

Figure 3.4 Quality of ownership is maximized by optimizing customer satisfaction, availability, and operating cost.

To do this, the design team must (1) develop or be given a clearly defined vision for the product (expectation standards); (2) gather customer needs, thoroughly understand them, and codify them into a well-defined set of customer needs; (3) develop and refine an easily measured and verified set of product specifications that effectively captures customer needs in engineering terms; (4) evaluate best in class competitor products with respect to the product specifications; and (5) develop a product concept that is superior to the competition with respect to each of the product specifications. Methods and practices for doing this are presented and discussed in Chapters 8 through 11.

Quality of Ownership

Quality of ownership relates to the experience the customer has as a result of owning and operating the product. Considerations such as ease of use, operating cost, reliability, serviceability, maintainability, condition of the product when purchased, and customer service all influence this quality. High quality of ownership is important because it is what causes a customer to become a repeat buyer and to become an advocate of the product in the marketplace. Also, this quality, more than any other, contributes to and sustains, over time, the firm's reputation for manufacturing and selling high quality products.

Quality of ownership depends on how satisfied the customer is with the product, how available the product is over its useful life, and how costly the product is to own and use. Customer satisfaction will be high if the product performs as it should and if it is perceived to be in excellent condition when purchased. A defect is anything that detracts from customer satisfaction. If a fastener is missing or the paint is blemished or a highly visible feature is not perfectly aligned, the customer's satisfaction with the product is diminished. Preventing defects from reaching the customer is obviously an important part of manufacturing quality control. Making it easy for manufacturing to do this is the job of product design. For example, by realizing that the defects mentioned above could occur, the design team can avoid potential customer dissatisfaction by developing a design which guards against the possibility of missing fasteners and blemished paint and by providing a guiding surface which insures correct alignment of the visible feature.

Improving customer satisfaction requires paying very close attention to detail. One approach for doing this is to perform in-depth assessments of the design at an early stage of the design, so that changes and improvements can be easily implemented. The goal of these assessments is to (1) identify all defects that could possibly occur, (2) determine the likelihood of a defect occurring and reaching the customer, and (3) implement corrective actions which prevent or reduce the probability of occurrence. A variety of quality assessment methods are available for doing this. Examples include cause-and-effect diagrams and failure mode and effects analysis (FMEA). The FMEA method is discussed in Chapter 23.

Availability is defined as the fraction of time a product (or system) is able to function. Product availability obviously depends on subsystem and component availability and is determined by the amount of time spent repairing an unplanned failure and the time required for routine or planned maintenance and service. Availability, A, is expressed mathematically as,

$$A = 1 - \left(\lambda T_R + \frac{T_M}{T} \right) \tag{3.5}$$

where λ = failure rate in failures/hour,

T_R = mean time to repair (MTTR) in hours,

T_M = average preventative maintenance time in hours/year,

T = 365 days/year x 24 hours/day = 8760 hours/year.

Equation (3.5) shows clearly what must be done to design a product for maximum availability. First, the product must be designed to have very low failure rates. This can be done by assessing the failures that might possibly occur using FMEA (see Chapter 23) or other techniques, performing extensive life testing to determine the mean time between failures (MTBF), and reviewing field experience gained from existing designs and then correcting the design as needed. Once the product goes into production, the philosophy of continuous improvement can be used to further reduce the failure rate and improve the product availability.

Assuming that all failures cannot be avoided at a reasonable cost, the second step for improving availability by design is to design the product so that it is very easy to repair. One way of doing this is to develop a product structure that makes it easy to remove and replace particular modules that are subject to wear out and failure (see Chapter 21). A second approach is to design the product so that it is easy to disassemble, repair, and reassemble with a minimum of adjustments and cost. Both of these strategies help to minimize the mean time to repair the product (T_R).

The third step is to minimize T_M by designing the product so that preventative maintenance is easy to perform and requires little time. In addition to minimizing λ , T_R, and T_M, availability can be further maximized by designing detectability into the product. Detectability is a characteristic of the product that makes it easy to monitor and/or check for component deterioration and the need for repair or preventative maintenance (Bloch and Geitner, 1990). Detectability enhances the quality of ownership by replacing the inconvenience, cost and time associated with unplanned failure with the more acceptable and less disruptive inconvenience of planned or preventative maintenance.

Operating cost is a measure of ease, safety, and economy of use. Depending on the product, these factors are likely to be included as customer requirements and are often determined as part of the quality of concept. In many ways, operating cost is a measure of "total product value" from the standpoint of using or operating the product. If the product is easy, safe, and inexpensive to use or operate, then it is a good design because quality of use is high and operating cost and time are low (see Eq. (3.1)).

Serviceability and durability are also important factors effecting operating cost. If a product is easy to service for low cost, operating cost (and inconvenience) is reduced. Designing the product for ease of assembly (Chapter 14) as well as planning for serviceability early in the design process are some ways to achieve a highly serviceable design. *Durability* is the amount of use the product can withstand before it is preferable to replace the product rather than to continue to repair it. Hence, a durable product is less expensive to own. Durability is often a function of the materials used and generally involves a trade-off between manufacturing cost and operating cost. Understanding customer needs and demands is therefore key in maximizing this quality.

Operational Robustness

A *robust design* is one that is insensitive to and/or tolerant of change and variation in hard to control variables that disturb or degrade the function of the product (called "noise" by Taguchi and Yuin (1979) and other authors). *Operational robustness* refers to the product's ability to tolerate variation in hard to control variables that influence the product function. These variables typically (1) are associated with the environment in which the product operates, (2) arise as the result of changes and degradations that occur over time and/or with product use, or (3) occur due to variation from product to product manufactured under the same specifications. Examples of environmental variables that influence the product during use include temperature, humidity, input voltage, dust, external load, time rate of load application, type of use, and so forth. Loss of strength due to corrosion, wear of mating parts, deterioration due to operation at elevated temperature, and shifts in calibration or adjustments exemplify the type of hard to control property or variable that influence product function and that change during use. Variations in part dimensions and calibration values that result from the manufacturing process are representative of product to product variation.

In developing an operationally robust design, the design team seeks to maximize functionality and minimize the effects caused by these disturbances. A truly robust product is one that has been consciously designed to be robust. One approach for maximizing robustness is to employ the following three-step process: (1) develop a robust conceptual design, (2) optimize the parameter settings, and (3) establish design specification widths. In step 1, the design team looks for an inherently robust design concept. Step 2 is aimed at minimizing the effect of variation. The goal is to make the design insensitive to hard to control disturbances and in this way increase the allowable design specification width. Actual values for the design specification widths (tolerances) are set in step 3 with the goal of making these widths as large as possible while still satisfying functional requirements.

3.3.2 Internal Quality

Internal quality implies product characteristics such as inherent ease of manufacture and assembly, consistent product characteristics from product to product, insensitivity to hard to control disturbances, and minimal scrap rates, rework, and warrantee claims. This quality is "internal" because the firm reaps the primary benefits and is of vital importance because it directly influences total cost of the product.

Internal quality depends on the degree to which the design and operating characteristics meet established standards, how easy the product is to manufacture and assemble, and how sensitive the design is to changes in

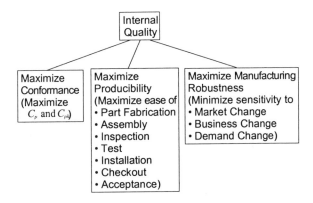

Figure 3.5 Internal quality is maximized by maximizing conformance, producibility, and manufacturing robustness.

demand, business needs, and product and manufacturing technology. In other words, it depends on the qualities of conformance, producibility, and manufacturing robustness (Fig. 3.5). What constitutes acceptable quality of these factors and how they can be maximized by design is discussed as follows.

Conformance

Conformance refers to how well the actual product or component characteristic conforms to design intent, where *design intent* is the intended value or target value of the characteristic. Conformance is acceptable when the actual value of a product or component characteristic is within the design specification width (Fig. 3.6a). Conformance improves as the actual value of a characteristic approaches design intent and is a maximum when the actual and target values are the same. Hence, the degree of conformance depends on both the amount of variation and on how close the actual value is to the target value. This corresponds with the Taguchi and Yuin (1979) view that any deviation from design intent causes a loss either to the customer or to the company (Fig. 3.6b).

Design specification widths (or tolerances) are necessary because, in general, no matter how much care and effort is expended to reproduce a characteristic identically from unit to unit, some variation occurs. In most situations, this hard to control variation is normally distributed about a mean value (Fig 3.6c). For normal distributions, variation is measured in standard deviations ($\hat{\sigma}$) from the mean. The normal variation, defined as process width, is $\pm 3\hat{\sigma}$ about the mean. The Capability Index (C_p) is used to quantify conformance, where:

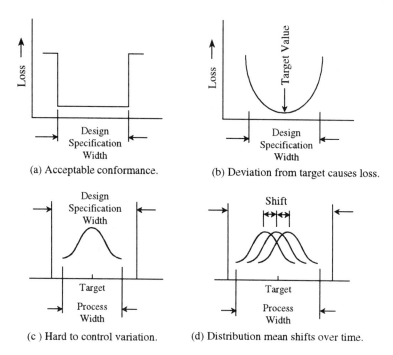

Figure 3.6 Conformance depends on the design specification width, the process width, and the location of the process mean with respect to the target value.

$$C_p = \frac{Design\ Specification\ Width}{Process\ Width} = \frac{Upper\ Limit - Lower\ Limit}{6\hat{\sigma}} \qquad (3.6)$$

Equation (3.6) is based on the assumption that the actual mean and design intent is the same, i.e., that the distribution is centered at the target value. Unfortunately, for most processes, the mean will shift over time with respect to design intent (Fig. 3.6d). For this more realistic case, the Capability Index is adjusted with a factor k, and becomes C_{pk}, where:

$$C_{pk} = C_p\ (1 - k) \qquad (3.7)$$

and

$$k = \frac{Process\ Shift}{(Design\ Specification\ Width)/2} \qquad (3.8)$$

As shown by Eqs. (3.6) through (3.8), conformance increases as C_p and C_{pk} become larger. Therefore, to improve the quality of conformance, the product should be designed in a way that makes it possible to use wide design specification widths and to use easily controlled and highly repeatable processes. Once the product is in production, conformance can be further improved by closely controlling production processes so that process standard deviation are small and means remain centered at design intent over time.

Producibility

Producibility relates to the ease with which the product can be manufactured, assembled, inspected, and tested. It also includes considerations such as availability of supplied components, raw materials, and production resources, and the clarity and simplicity of the detailed design information used to manufacture and support the product. A producible design is one in which the product design and associated production planning is appropriate for the quantity required, trade-offs are possible to achieve the optimum of the least possible cost and the minimum time, and conformance to established specifications is acceptable (Military Handbook 727, 1984).

A high degree of producibility increases internal quality by reducing quality risks, manufacturing complexity, hard to control randomness, production uncertainty, and life cycle support complexity. Producibility is maximized by (1) identifying design concepts that are inherently easy to manufacture for the best balance of cost and time, (2) focusing on component design for ease of manufacture and assembly, and (3) integrating the design of the product and manufacturing process to ensure the best matching of needs and requirements.

Manufacturing Robustness

Manufacturing robustness is the ability of the manufacturing system to tolerate product design changes and production volume (demand) changes that occur due to changing market needs, business needs, and technology innovation. Examples include product features, performance specifications, product styling, and supplier/part/material availability. The objective is to minimize capital investment and timing consequences incurred due to inevitable and necessary product design change. Manufacturing robustness is a very subtle quality of a product that can have a far-reaching effect on long-term profitability of the firm. A product that has a high degree of manufacturing robustness can be adapted quickly to changing market and business needs with a minimum impact on manufacturing operations and investment.

To improve the manufacturing robustness of the product, the design team must look five or six product generations into the future and anticipate changes that are likely to occur and then consciously plan for them in the design. One approach for doing this is to utilize the following three-step procedure: (1) determine how the product and manufacturing method might change, (2) develop alternative design concepts and manufacturing methods that will accommodate anticipated change, and (3) evaluate the alternatives and select the alternative with the greatest potential for maximizing total product value. What product and manufacturing changes might occur and how the product could be designed to accommodate these changes can be investigated by evaluating the product design with respect to the following questions:

- How might the product styling and functionality change over time? How might customer needs change? What applicable new technologies are likely to become available? How could the product be designed to accommodate these changes?
- How might the manufacturing method or production technology change over time? How could the product be designed to accommodate these changes?
- What product or model variations are planned? How does the product and manufacturing method accommodate these variations? What new variations may be required in the future? How can the product and manufacturing method be designed to accommodate these changes?

Each of these questions should be answered in as much detail as possible, given the product and process definition available. If the product design and manufacturing plan are fairly complete, then the questions will help identify vulnerabilities to change. If the product and manufacturing plan is less well developed, then the questions will help point the way to a more manufacturing robust design solution.

3.4 TOTAL COST

All design decisions, both large and small, impact product total cost in one way or another. Traditionally, only direct manufacturing costs have been calculated when evaluating the goodness or acceptability of a design decision. *Direct cost* is typically computed as the sum of the variable and fixed costs, that is,

$$Direct\ Cost = V\left(m+l+p\right)+F \tag{3.9}$$

where V, m, l, p, and F are as defined previously in Section 3.1. Evaluating design decisions based on direct cost is convenient because these costs can be determined and tracked relatively easily. Unfortunately, Eq. (3.9) neglects the indirect cost incurred by the firm in manufacturing and marketing the product.

Indirect cost includes all those costs such as overhead and administrative costs, design cost (time and equipment), and systems "debugging" and problem solving costs, which can't be easily attributed to any one aspect of the product and are therefore lumped together and accepted as the "cost of doing business." In design for profitability, we recognize that design decisions impact both direct and indirect cost. Most importantly, indirect cost can often far outweigh direct cost, especially when computed over the life cycle and lifetime of the product. Therefore, to develop a profitable product design, the cost implications of design decisions need to be evaluated in terms of total cost.

Design related indirect cost is subdivided into three components: systems cost, S; development cost, D; and time cost, T. *Systems cost* includes all of the day-to-day indirect operations and overhead cost associated with manufacture, distribution, marketing, service, and disposal of the product. Examples include costs associated with purchasing, supplier relations, order processing, inspection, material handling, receiving, shipping, service parts, catalogs, promotional material, worker training, and so forth. *Development cost* includes the direct cost of designing the product and launching it into production plus all of the overhead and equipment cost associated with CAD/CAM/CAE systems, testing labs, prototype shop, R&D labs, and other design support systems that facilitate the product realization process. Finally, *time cost* can be thought of as the cost of lost sales and market share incurred as a result of being late to market or second in the market with a new product. Adding these costs to the direct cost of Eq. (3.9) results in the expression for total cost given by Eq. (3.3). Ideally, we would like to click an icon in a CAD system and receive immediate feedback on the total cost consequence of a design decision. Unfortunately, accurate total cost models and the cost structures required to develop them do not exist in most companies. As mentioned previously, this is an over arching concern that is addressed throughout this book.

3.5 TOTAL TIME

Each design decision, both large and small, contributes in some way to total time. This is illustrated by Table 3.1, which lists some of the many components of total time. Taking time out of the new product introduction cycle or order-to-delivery cycle or product servicing cycle can yield numerous benefits. Customer satisfaction is improved because short cycles give the customer what he or she wants in a timely manner. Short cycles also reduce total cost because, to achieve short cycles, waste and nonvalue activity must be eliminated. There are many design-related strategies that can be employed to help reduce total time. In the following, we briefly overview some of the most important of these.

Table 3.1 Some Major Components of Total Time

Product Design Cycle	Order-to-Delivery Cycle	Operational Life
Design & Dev. Time	Order Processing Time	MTBF
CAD/CAM Time	Manufacturing Lead Time	MTTR
Prototype Fabrication	Piece Part Cycle Time	MTTF
Test/Validation/Fix	Inspect/Test/Calibrate Time	Service Time
Tool Design/Prove-Out	Setup/Tool Change Time	Service Interval
Launch/Ramp-Up	Material Handling Time	
	Assembly Time	
	Delivery/Installation Time	

1. **Re-Engineer the Design Process.** Make organizational and procedural changes that enhance communication between functions, encourage design for manufacture, and simplify and optimize workflow. Use concurrent engineering practices and the team approach. If appropriate, co-locate design and manufacturing engineering. Use technologies such as teleconferencing, e-mail, and the Internet to overcome wide geographical separations. Carefully define the product realization process so that the design team knows exactly what it must do during each step of the process. Institute design reviews to ensure economic viability of design projects and to facilitate simultaneous achievement of product manufacturability and tight schedule commitments.

2. **Use Integrated Design Systems.** Integrate the CAD/CAM/CAE system so that product and manufacturing engineers work on compatible systems that share information in a seamless environment. Employ computer-aided design tools and methods as an integral part of the product realization process. Instead of computerizing manual practices such as drafting, use the computer to eliminate these practices where possible and to replace them with new computer-aided practices that reduce design time and effort by fully utilizing the power of the computer. For example, parametric and feature-based CAD/CAM tools now enable design changes to be made and then propagated through the model with a few high-level commands. Using these tools, engineering analysis, manufacturing, and inspection can be automated and integrated with design to a higher degree than ever before.

3. **Utilize Computational Design.** Many product developments rely on experimental or "test and fix" procedures involving costly and time consuming construction and testing of many physical prototypes. Eliminate these practices by developing a "science base" for the product technology involved. Once an appropriate science base is available, computer

simulations and other analysis techniques can be used to more quickly select the best starting point for a new design.

4. **Standardize Where Practical.** Reduce design time by leveraging commonalties present in different product models and variants. Do this by developing product architectures that facilitate the use of modular and standardized subsystems and building block parts. Rationalize the variety of choices available for purchased components such as threaded fasteners and ball bearings. New products are then quickly designed since only the unique, nonstandard aspects of the new product must be designed. Time required for component testing, vendor selection and qualification, and a myriad of other design-related tasks is also reduced.

5. **Streamline Order Processing.** Use the computer to take time out of the order fulfillment process. Develop computer software that enables a sales representative to negotiate and complete a purchase order or release at the customer's place of business and then insert the order into the plant's production schedule via modem and confirm it before leaving the customer. If the product needs to be customized to satisfy specific customer requirements, develop software that processes the specific customer requirements and returns a set of approval documents via modem to the sales representative for immediate review and acceptance by the customer. The best solutions to this type of order engineering scenario are generally realized when modular designs or designs which allow customization at the end of the production line are developed with a particular order engineering approach in mind. Such designs are excellent examples of how marketing and manufacturing strategies coupled with a well-managed design-marketing interface can create significant competitive advantage for a company.

6. **Design to Reduce Manufacturing Lead-Time.** *Manufacturing lead-time* (MLT) is the time required to process the product through the plant. A short MLT implies less manufacturing time and effort and therefore less cost. Also, the shorter the MLT, the sooner the product can be sold and the company reimbursed for its investment in raw material and labor. Most importantly, a short MLT means the customer gets the product when needed. The MLT can be reduced as part of the product design in two ways: (1) use design for manufacture and assembly practices, and (2) coordinate the product and manufacturing process design. Groover (1987) calculates MLT as

$$MLT = \sum_{i=1}^{n_m} \left(T_{su} + QT_o + T_{no} \right)_i \qquad (3.10)$$

Table 3.2 Strategies for Reducing Manufacturing Lead Time (Groover, 1987)

Strategy	Effect*
1. Specialization of operations	Reduce T_o
2. Combined operations	Reduce n_m, T_h, T_{no}
3. Simultaneous operations	Reduce n_m, T_o, T_h, T_{no}
4. Integration of operations	Reduce n_m, T_h, T_{no}
5. Increase flexibility	Reduce T_{su}, WIP; increase U
6. Improve material handling and storage	Reduce T_{no}, WIP
7. On-line inspection	Reduce T_{no}, q
8. Process control and optimization	Reduce T_o, q
9. Plant operation control	Reduce T_{no} ; increase U
10. Computer-integrated manufacturing	Reduce design time, production planning time; increase U

* WIP - work in process; q - scrap rate; U - equipment utilization.

where n_m is the number of machines or operations and T_{su}, Q, T_o, and T_{no} are the setup time, batch quantity, operation time, and nonoperation time for each machine and/or operation, respectively. Nonoperation time includes handling, storage, inspections, and other non-value added activities. Table 3.2 presents a summary list of 10 strategies that are commonly employed to reduce MLT. Designing the product and the manufacturing process as a coordinated system often facilitates these strategies.

3.6 KEY TAKEAWAYS

Design for profitability seeks to maximize those qualities of the product that delight and satisfy the end customer while simultaneously minimizing total cost of the product. This requires an "external" perspective that focuses on creating a functionally and aesthetically desirable product that satisfies customer needs. It also requires an "internal" perspective that focuses on decomposing the product into designed and standard parts and arranging them in a product structure that minimizes total cost over the long term.

- Profitability is a function of total product value, which is in turn a function of total product quality, total cost, and total time. By designing to maximize total product value, the team maximizes profitability.

- Each design decision, both large and small, contributes in some way to total product value. Good design decisions increase total product value by increasing total quality and/or decreasing total cost and total time.
- Total product quality is the totality of features and characteristics of a product that bear on its ability to satisfy stated or implied needs. The design of the product is the pivotal starting point for achieving total product quality. A good product design not only contributes directly to total product quality, it also makes manufacture and life cycle support of the product easier. This in turn helps create and foster a more positive and proactive company-wide attitude toward quality.
- To design for total product quality, the design team must systematically focus on quality as a design objective. One way to do this is to view total quality in terms of components of external and internal quality. Total quality is then maximized by maximizing the factors that contribute to each component.
- Total cost includes all costs, both direct and indirect, that occur over the product life cycle. The impact of individual design decisions on total cost is often very difficult to predict especially in the early stages of design. Designing to minimize total cost is therefore a major challenge of good design.
- Total time is the composite of all times that are effected or determined by product design decisions. Good product designs are quickly designed, require a short manufacturing lead-time to produce, and can be repaired and serviced quickly. Reducing total time is a primary focus of good design because it both increases customer satisfaction and decreases cost.

4

Improving Early Design Decisions

4.1 INTRODUCTION

Design decisions made during the early phases of design are especially critical because they have a tremendous impact on the total cost of the product (Fig. 4.1). Often, high quality design decisions made during the early stages of design can equal years of cost reduction and design improvement made after design release. It is therefore imperative that early design decisions be well thought out and carefully made. Critical early design decisions typically fall into two categories: (1) the identification and selection of the best working principle and (2) the identification and selection of the best part decomposition for the product. In this chapter we examine the nature of early design decisions and then present a set of best practices that provide a philosophical framework for improving the quality of both of these categories of design decisions.

4.2 ENGINEERING DESIGN PROCESS

Early design decisions are those that involve the initial definition of the product's design. These decisions are generally made during the *engineering design process*, which typically involves the following design activities:

1. Clarify and define the requirements of the product or design.
2. Develop a working principle or physical concept for fulfilling required product functions.
3. Decompose the physical concept into subassemblies and components; determine the geometric arrangement (layout) of components; establish dimensional relationships between components.
4. Decide which components are standard and which must be designed.

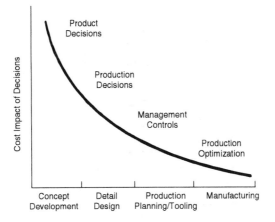

Figure 4.1 Cost impact of decisions over the product life cycle. Early design decisions effect life cycle cost far more than years of manufacturing improvement performed after the design is released.

5. Select the general type of material (e.g., polymer, metal) and basic manufacturing process (e.g., casting, machining) to be used for each designed component, if not already determined.
6. Determine the configuration (i.e., size, shape, external and internal geometric features) of each designed component.
7. Select a specific material and manufacturing process for each designed component.
8. Establish dimensions and tolerances for each designed component.
9. Supply additional dimensions, tolerances, and detailed information required for manufacture and assembly of the components.

These activities proceed from the general to the specific and are typically performed in the order listed when developing a new product design.

The process begins by conceiving a physical concept for the product based on customer needs and a product specification and then creating a preliminary layout of the design that embodies the physical concept (Fig. 4.2). This initial phase is often referred to as the conceptual design stage and typically involves activities 1 through 5 listed above. The preliminary layout represents a conceptual arrangement of parts that implement the physical concept and working principle of the product. It is preliminary because, at most, only key dimensions and relationships between parts have been specified; the actual size, shape, and detail features of the parts are, as yet, either undefined or only partially defined.

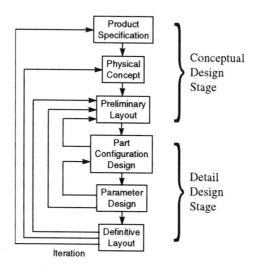

Figure 4.2 Iterative model of the engineering design process.

The preliminary layout is then developed into a completed design by developing a configuration design and parametric design for each designed component. Configuration design involves determining the general size, shape, and detail features of the designed components (activities 5 and 6), while parametric design involves assigning specific material properties, dimensions and tolerances (activities 7, 8, and 9). Collectively, these design activities are often referred to as the detail design stage. During detail design, the preliminary layout changes and evolves iteratively as questions are answered and uncertainties resolved. The end result is the definitive layout or final design. The definitive layout contains the design information required to fabricate and assemble the parts.

4.3 CONCEPTUAL DESIGN

Conceptual design involves the identification and selection of the best working principle and part decomposition for the product. To illustrate the importance of early, conceptual design decisions, consider the design of a simple compressor for use in a vapor compression refrigeration cycle. Given rotary input motion, the working fluid may be compressed using a slider-crank mechanism to move a piston back and forth in a bore or by a vane assembly rotating in a cylindrical cavity. Assuming both design alternatives develop the pressure and flow rate required, which is the better choice?

To answer this question, we imagine the different ways each concept could be decomposed into parts and then compare the best decomposition for each alternative. Doing this, we find that the rotating vane concept has less rubbing (wear) interfaces between parts and fewer theoretical parts (parts that must be separate for fundamental reasons), and can be decomposed into parts which have simpler, easier to manufacture geometry. Also, dimensional tolerances for the rotating vane concept are, in general, less critical because sealing and durability are not as dependent on maintaining close fits between parts. Hence, when the working principle and part decomposition are considered together, the rotating vane concept is seen to be the better choice because it requires fewer parts which are easier to fabricate. By selecting the rotating vane concept and then making design decisions that result in easy to manufacture and assemble parts, the design team ensures long-term profitability because total cost of the rotating vane concept is intrinsically lower than the reciprocating piston concept.

This simple example illustrates three critical aspects of conceptual design:

1. For most products, many different physical concepts are possible. By *physical concept*, we mean the working principle or solution concept employed to provide the desired functionality of the product. The physical concept embodies the way in which the product performs or provides its intended function. The reciprocating piston alternative and the rotating vane alternative for compressing the working fluid in the above example represent two different physical concepts. The key to achieving a high external quality often lies in identifying and selecting the right (or best) physical concept for the product. This fact, which is well known, is a primary motivation for creativity and innovation in product design. It is also the motivation for the research and development activities conducted by many manufacturing firms.

2. Just as important, but not as clearly recognized, is the fact that there are usually many different ways in which a given physical concept can be divided or decomposed into parts. We refer to this important aspect of conceptual design as *part decomposition*. For many products, identifying and selecting the right part decomposition can be as important or even more important than the physical concept itself. Part decomposition determines the ease of assembly, testability, and serviceability of the product. It also determines the number and complexity of the designed parts, which in turn, influence material and manufacturing process selection, tooling cost, and a myriad of other factors. For many products, part decomposition, more than any other early design decision determines total cost and internal quality of the design.

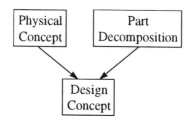

Figure 4.3 Design concept = physical concept + part decomposition.

3. The physical concept and part decomposition together define the
    design concept (Fig. 4.3). It is the combination of physical concept
    and parts decomposition that determine (often synergistically) many
    quality related product characteristics such as robustness, reliability,
    maintainability, durability, and operating efficiency. For many
    products, therefore, the selection and development of the design
    concept is the key to achieving high total product value.

   In most product design situations, the opportunity for selecting the physical
concept for a product seldom occurs more than once in the products lifetime.
This is because this choice establishes many irreversible properties and
characteristics of the product and its method of manufacture. Once the
customer becomes accustomed to the product and its method of operation
and/or significant investments in tooling and equipment have been made and/or
regulatory approvals have been gained, it is very difficult to make extensive
changes to the physical concept. Fortunately, in many mature products, it is
possible to substantially change the design concept of the product without
affecting the physical concept by changing the way the product is decomposed
into parts. When this is possible, manufacturability and quality of the product
can often be significantly improved in a redesign effort without adversely
affecting customer perception and regulatory approvals.

4.4 DETAIL DESIGN

Early design decisions that determine component configuration and parametric
design must also be carefully considered and evaluated in terms of total cost.
Traditionally, detail design decisions focus primarily on piece-part cost. This is
because tooling cost and piece-part production rates are often established by the
component configuration design while process cycle time as well as component
weight, failure modes, and cost are determined by the parametric design. For
these reasons, it is not unusual for cost targets to be ultimately achieved by

paying close attention to piece-part design and striving to control and minimize piece-part cost.

It is important to note, however, that piece-part cost is just one component of total cost. It is therefore possible that particular part decompositions might result in higher piece-part cost, but lower total cost. This is because a well conceived part decomposition not only allows and facilitates part configurations which are easily designed and tooled, it also impacts assembly cost, cost due to quality risk, cost of non-value added manufacturing activities such as material handling, and it leverages the benefits of standardization, building block parts, and product families. Often, a well formulated part decomposition facilitates economies of scope and scale which are impossible when each product component is optimized for minimum cost as a unique part fabricated from a unique starting material.

4.5 BEST PRACTICES

We believe the following practices constitute the essential success factors for making high quality early design decisions. When the conceptual and detail design of a product is developed using these practices, the likelihood of long-term product success and profitability is substantially increased.

1. **Use a Disciplined Design Process.** A disciplined design process includes the use of the team approach, formal design reviews, and best practice guidelines. The team approach makes needed design information available to support each design decision and helps ensure that the best possible decisions from every standpoint are made. Formal design reviews help insure optimal design by providing an independent check to be sure all customer needs are satisfied and that the design will yield a satisfactory profit. Best practice guidelines provide the guidance, checklists, structured methods, and tools needed to ensure a systematic and informed design approach.

2. **Develop a Design Strategy.** A clearly articulated design strategy helps eliminate and narrow choices, makes the best choices more obvious, and constrains choices that would otherwise be arbitrary. The design strategy can be based on many different considerations. For example, a process driven design strategy would cause the product to be designed in a way that allows a particular assembly sequence or method of manufacture to be used. A product family strategy, on the other hand, might cause the product to be designed to allow the use of "building block" components or other standardized modules or practices. Often, a well-defined design strategy is the single most important ingredient for improving early design decisions.

3. **Develop a Comprehensive Set of Customer Needs.** Good engineering design depends on correct technical trade-offs, product innovation, and a deep commitment to developing a functionally and aesthetically appealing product that delights and satisfies the customer. This requires a comprehensive consideration of both external and internal customer needs. *External customer needs* typically include the needs of those who distribute, buy, use, and service the product. *Internal customer needs*, on the other hand, are those that relate to business issues and strategies, manufacturing, and life cycle product support. For some design situations, understanding the problem of design involves identifying external customer needs and developing an appropriate product specification. In other situations, the internal manufacturing needs of a product family or the manufacturing constraints of an existing facility must be understood. To develop a good design, all customer needs, both external and internal, must be identified and properly characterized.

4. **Generate Many Alternative Physical Concepts.** The likelihood of identifying and selecting the best physical concept is greatly increased when many different alternative concepts are proposed and systematically considered.

5. **Generate Many Alternative Part Decompositions.** The likelihood of identifying and selecting the most appropriate part decomposition for ease of assembly, ease of component manufacture, ease of service, and high total product quality is greatly increased when many alternative part decompositions are proposed and systematically considered.

6. **Base Selection Decisions on a Disciplined Evaluation of Alternatives.** The selection of a particular physical concept and part decomposition defines the design concept and irreversibly establishes many of the characteristics of the product. In most product designs, the design concept can have a strong and far-reaching influence on market success and on total cost, time, and quality. The use of an explicit and disciplined selection process helps insure objectivity and customer focus. It also helps guide the team through the decision making process, documents the process, and often provides additional insight into ways to further improve the design.

7. **Avoid Undesirable Interactions.** If a vehicle pulls to the left or to the right when the brakes are applied, an undesirable interaction is said to exist between the steering and braking functions of the vehicle. Similarly, if a critical dimension depends on the assembly of several components, an undesirable interaction exists when high cost precision parts and extreme care during assembly are required to

ensure achievement of the critical dimension. Undesirable interactions between product functions or product components lead to poor performance, high cost, and low quality. Systematic evaluation of the design concept aimed at identifying and eliminating undesirable interactions therefore ensures good design. For example, by changing the faulty part decomposition discussed above so that the critical dimension is isolated within a single component, the undesirable interaction between assembled components is avoided.

8. **Use a "Just Build It" Philosophy.** The best and quickest way to verify a physical concept or part decomposition idea is to build a simple physical model. Simple mockups and models quickly resolve design uncertainties and provide insight into the implications of design decisions that would be very difficult to obtain in any other way. Early experimentation with hardware embodiments of ideas and concepts can often surface "unasked" questions and identify latent needs. It can also provide important insight into approaches for decomposing the physical concept into easy to manufacture and assemble components. We call this experimental approach to design decision making the "just build it" philosophy of conceptual design. The tendency of the modern engineer is to resolve uncertainty using engineering analysis. Analytical approaches can be time consuming, however, and often requires more detail design information than is available during conceptual design. The "just build it" philosophy avoids the need for detail design information, saves valuable time by resolving uncertainty quickly and unambiguously, and provides insights that can't be obtained in any other way.

9. **Compare with "Best in Class" Competitor Products.** Long-term market success requires that the product be superior to the competition with respect to all critical attributes such as functionality and performance, cost, and quality. Benchmarking the competition and "reverse engineering" best in class competitive products provides the information needed to make design decisions that lead to superiority.

10. **Maintain a "What-If" Mentality.** The time to question design decisions and to explore all possibilities and options is in the early conceptual stages of design when maneuvering space is wide and hardware is still remote. A "what-if" mentality constantly strives to find a better way, seeks to avoid undesirable interactions and downstream consequences, and maintains openness to rethinking and reconsidering all decisions.

4.6 KEY TAKEAWAYS

The quality of early design decisions often determines the long-term success of a new product development project. Making high quality early design decisions requires early and systematic consideration of all life cycle needs of a product. The objective is to define the best design concept as quickly and efficiently as possible by avoiding lengthy and costly design iterations.

- Early design decisions determine the design concept for the product. The design concept converts a marketing concept or idea into designed parts and standard components that assemble and work together to provide the desired product function.
- The design concept defines how the product provides required functions (physical concept) and how the product is to be divided into designed and standard components (part decomposition). These decisions are often the most critical from the standpoint of total cost and customer satisfaction.
- Making quality early design decisions involves (1) identifying the best physical concept for satisfying customer requirements, (2) identifying the most appropriate part decomposition for ease of manufacture, assembly, and life cycle support, and (3) integrating these into a design concept that optimizes total product value.
- The likelihood of developing a good product design is greatly enhanced when proven best practices are employed.

5

Principles of Good Design

5.1 INTRODUCTION

Are there rules for good design and if so, what are they and how should they be applied? How can the goodness of a design decision be judged? If there is a problem with a particular design solution, is there a way of analyzing the problem that will reveal the underlying cause and point the way to the best redesign? In this chapter, we present underlying principles of good design that can be used to answer these questions and others like them.

N.P. Suh and his associates at MIT have shown that good design embodies two basic principles (Suh, Bell, and Gossard, 1978). Yasuhara and Suh (1980) formally stated these principles as design axioms and corollaries. The theory is further developed by Suh (1990). Unfortunately, the axioms and corollaries are relatively abstract and difficult to apply by practicing designers. Our approach is to recast the axioms as simple design rules and to illustrate their use by discussing a variety of practical design examples. The examples are also used to derive and illustrate the corollaries where appropriate.

5.2 PRINCIPLES

The first principle is that undesirable interactions between various functional requirements of a product should be avoided. The second basic principle is that good designs maximize simplicity, that is, they provide the required functions with minimal complexity. These principles can be formalized as follows:

1. In good design, undesirable interactions are avoided.
2. Among the designs that avoid undesirable interactions, the best design is the one that has the minimum information content.

Our contention is that the root cause of many quality, manufacturing, and performance problems can be traced to undesirable interactions between various components or systems of a product design. *Undesirable interactions* occur when functional requirements become coupled in undesirable ways. For example, the steering function and the braking function of an automobile interact undesirably when application of the brakes causes the vehicle to pull to the left or right.

Similarly, we contend that cost and quality of a product are directly driven by the complexity of the design. Therefore, a good design will be as simple as possible. *Information content* is a measure of complexity. By minimizing the information content of a design, the simplicity of the design is maximized. One way to think of information content is to imagine the set of instructions that would be required to explain every aspect and detail of the product's manufacture, operation, use, maintenance, and disposal. Consider, for example, the number of separate activities and the number of instructions per activity required to manufacture a particular design. The best design would be the alternative requiring the least number of activities with the fewest instructions per activity. For instance, a product composed of few parts would require less manufacturing and assembly activities than one composed of many parts. A part designed so that it is machined on only one surface using one setup would require fewer instructions to fabricate compared to one having machined features on several surfaces, each requiring a different setup and machining operation. A dimension that is consistently at its target value from part to part would require fewer instructions to deal with than one that varies widely or is consistently off target.

5.3 METHODOLOGY

The methodology for avoiding undesirable interactions and selecting design alternatives that minimize information content is essentially one of using a design-analyze-redesign process to home in on the best design (Fig. 5.1). The process consists of three steps:

1. **Understand:** analyze the proposed design solution for undesirable interactions and extra information content.
2. **Create:** propose redesigns that avoid the undesirable interactions.
3. **Refine:** analyze the redesign alternatives and select the one having the least information content.

Excessive or difficult to solve quality, manufacturing, and performance problems are clear and loud signals that undesirable interactions may be present. Similarly, complex parts, excessive part counts, many adjustments

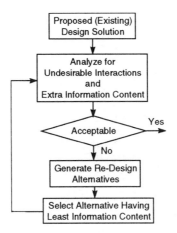

Figure 5.1 Design-analyze-redesign method for applying the design principles.

during assembly, many different materials or stock sizes, many separate or different fasteners, many different manufacturing operations and setups, many re-orientations, process uncertainty, and randomness all signal that the design is too complex. Avoiding undesirable interactions and excessive information content requires constant questioning and evaluation of each design decision. How could an undesirable interaction occur? What would be the symptoms? How could the part count be reduced? How can this adjustment be eliminated? Can randomness be reduced? "What-if" we did this? "What-if" this condition arose, what would happen?

The following examples illustrate how undesirable interactions can be identified and then eliminated by redesign. The examples also illustrate how the minimum information principle can be used to select and improve the best alternative among those alternatives that avoid undesirable interactions.

5.4 PRODUCT/PROCESS INTERACTION

Most undesirable product/process interactions occur because of deficiencies in the part decomposition selected. To illustrate, consider the design of a simple electrical circuit breaker (Fig. 5.2). The basic function of the circuit breaker is relatively simple. When an over current or fault on the line is detected, the armature is released and a spring force causes the armature to rotate about a pivot and, in so doing, open (break) the circuit. The basic elements involved in this action are shown for three different designs (part decompositions). Note that a cover plate (not shown), which is similar in shape and form to the base plate, attaches to the base plate and holds the assembly together.

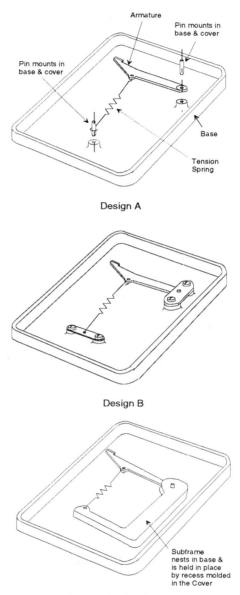

Figure 5.2 Spring-loaded circuit breaker mechanism (cover and details of mechanism not shown).

Design A is the original design. As can be seen in the figure, one end of a tension spring is attached to the armature and the other end to a pin that is supported by the base and cover plates. Similarly, the pivot about which the armature rotates when the breaker is tripped, is formed by a second pin that is mounted in and located by the base and cover plate. The assembly of the device is made very difficult because of the force exerted by the spring and the need to locate the pins in both the base and the cover. As a result, assembly must be performed manually with the assistance of complex assembly aids and tooling. Also, the process is extremely difficult to perform because of the force bias on the pin position caused by the spring force and the "mouse trap" nature of the resulting assembly. Because of the high production volumes involved, this difficult manual assembly is costly in terms of both time and money. And, because of the quality risk associated with the assembly process, each unit must be specially tested after assembly to insure proper operation.

In looking at design A, we see that the problem is caused by a *coupling* between the "enclose" functional requirement provided by the base and cover plates, and the "cause to rotate" functional requirement provided by the tension spring. The consequence is an undesirable interaction between the spring force and the assembly process. To eliminate the undesirable interaction, we need to *decouple* the functional requirements. Two possible alternatives, designs B and C, are proposed (Fig. 5.2). Design B eliminates the interaction by integrating the "cause to rotate" functional requirement into the base plate. Alternatively, design C eliminates the interaction by independently satisfying the "cause to rotate" functional requirement using a "subframe."

Both redesign proposals avoid the undesirable interaction identified in the original design. We now ask which part decomposition proposal, design B or C should be selected? The answer is provided by the minimum information principal. Comparing the information content of the two alternatives, we see that design C involves the addition of only one new part, the subframe, while design B involves several new parts. Also, the subframe allows the breaker mechanism to be assembled independently of the base and cover plate and avoids the need to install additional fasteners. Hence, design C requires less information content and should be selected. Because it has less information content, design C has fewer quality risks and can be assembled using faster, less costly assembly automation.

The following corollary to the principles of good design can be derived from the circuit breaker example:

Avoid design solutions in which functions are coupled. Decouple or separate parts or aspects of the design if functions are coupled or become coupled in undesirable ways.

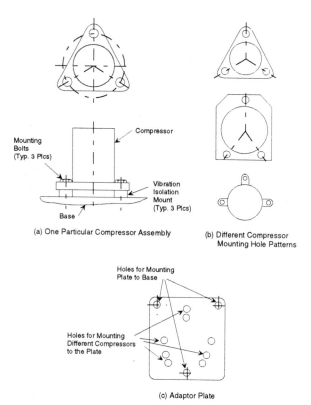

Figure 5.3 Compressor mounting design.

5.5 PRODUCT VARIETY

Undesirable interactions can arise due to purchased component variety. Consider the compressor and base plate assembly for a refrigeration product (Fig. 5.3a). In this particular part decomposition, the compressor, which is purchased as a supplied part, is mounted to the base plate via vibration isolation mounts at three points as shown. To control cost, it is occasionally necessary to change compressor suppliers. The undesirable interaction arises because the mounting bolt pattern is different for compressors supplied by different compressor manufacturers (Fig. 5.3b). As a result, different base plate designs and associated tooling must be maintained and supported to facilitate interchangeability of the different compressors. In addition, because the compressor is installed by a robot during assembly of the product, end of arm tooling and robot programming must be changed for each different compressor.

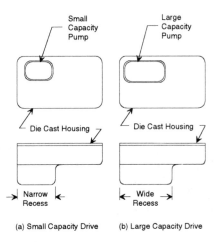

(a) Small Capacity Drive   (b) Large Capacity Drive

Figure 5.4 Schematic of a die cast housing assembly for a hydrostatic drive. The major difference between product variants is the size of the pump.

As is often the case in these situations, adding an adapter plate that interfaces the supplied component to the product can decouple this undesirable interaction. The adapter plate is mounted in the base via vibration isolation mounts using a standard, unchanging hole pattern (Fig. 5.3c). Installation of different compressors is facilitated by providing hole patterns in the adapter plate that match each compressor variant. Gripping the adapter plate instead of the compressor itself facilitates invariant robotic assembly.

In this example, use of the adapter plate effectively eliminates the undesirable interaction between the "compressor mounting" functional requirement and the "supplier variety" functional requirement. Also, replacing several different base plate and end of arm tooling configurations with one standardized configuration reduces information content. In addition to illustrating the "decouple" corollary, this example also illustrates the following important corollary to the design principles:

Use standard or interchangeable parts whenever possible.

Undesirable interactions can also arise due to variations from model to model in a product family. Consider, for example, the hydrostatic transmission pump and die cast housing assembly shown in Fig. 5.4. Several different models of the transmission are produced, each having a different power capacity. The primary difference between each model is the size of the pump. In the part decomposition shown, the pump is mounted in a recess formed by the die cast housing. Because the pumps are of different sizes, each hydrostatic transmission model requires its own unique die cast housing. The necessity to design and manufacture a different housing for each model indicates that an undesirable interaction exists between the "enclose," "support," and "power capacity" functional requirements. In this particular case, the undesirable interaction has resulted in the creation of an additional functional requirement, that is, "specialize the enclosure for each power capacity."

This undesirable interaction can be avoided by redesigning the transmission to make mounting and support of the pump independent of the die cast housing. If this is done in a way which does not introduce other undesirable interactions or compromise product functionality, then one die cast housing can be used interchangeably among all product variants, greatly reducing both information content and total cost. This example illustrates another important design corollary:

Minimize the number of design functions and constraints.

5.6 FORCE FLOW/GEOMETRY INTERACTIONS

Force flow/geometry interactions arise frequently in product and equipment applications. Force flow is visualized as lines of force passing through the member (Fig. 5.5a). The concentration of lines of force represents stress. The addition of a groove (Fig. 5.5b) interacts with the force flow causing the lines of force to bunch together. The result is a region of high stress or stress concentration. The maximum stress in the region of the stress concentration can be significantly higher than the average stress for the region (Fig. 5.5c).

Whether the presence of geometrical stress concentration represents an undesirable interaction between the force flow and component geometry depends on the nature of the material and type of loading involved. For materials permeated with internal discontinuities, such as gray cast iron, stress raisers usually have little effect, regardless of the nature of the loading (Juvinal, 1983). This is because geometric irregularities seldom cause more severe stress concentration than that already associated with the internal irregularities. If the loading is static (constant) and the material is ductile (i.e., the material deforms plastically when the stress exceeds its yield strength), then local yielding and subsequent redistribution of stress will null out any harmful effect of the stress concentration and it is customary to ignore stress concentration.

If, on the other hand, the material is both brittle and relatively homogeneous, or if special conditions exist (e.g., low temperature) that make a ductile material behave in a brittle manner, then stress concentration will weaken the part. This is true, even under static loading conditions, because

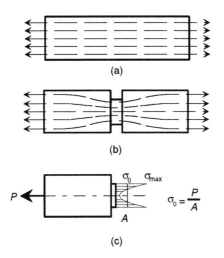

(a)

(b)

(c)

Figure 5.5 Force flow/geometry interaction. (a) Straight shaft; (b) shaft with a groove; (c) stress distribution on groove cross section.

brittle materials fracture in the region of highest stress when the stress exceeds the ultimate strength of the material. For *fluctuating (fatigue)* and *impact loading* of most materials, the interaction is also very undesirable and stress concentration must be avoided if possible.

Stress concentration can also have serious consequences in manufacturing processes. In solidification processes such as casting and plastic injection molding, forces developed during thermal contraction of the part can result in high residual stresses being frozen into the solidified part. Also, fracture or other defects can occur when the maximum stress exceeds the hot strength of the material. Similarly, tears and other defects can occur in regions of stress concentration in processes that involve large material flows such as sheet metal working and forging.

Avoiding undesirable interactions between the force flow in a component and the geometry of the component is therefore an important consideration in developing the best part decomposition. Design practices that help minimize the magnitude and effect of stress concentration regardless of the material or type of loading include the following:

1.   Avoid sudden changes of direction in the flow-lines of force.
     •   Avoid abrupt changes of cross section.
     •   Provide generous smoothing contours.
     •   Use large radii at fillets, grooves, holes, etc.

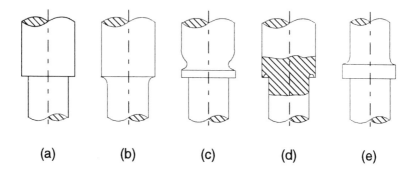

Figure 5.6 Reducing stress concentration in a stepped shaft. (a) Severe stress concentration; (b) use large radius if possible; (c) add groove if large radius is not possible; (d) undercut shoulder helps if modifications (b) or (c) cannot be used; (e) a short step might also help.

2. When there is a choice in the placement of a stress raiser, it should be located in a region of low nominal stress.
3. When the severity of the geometrical feature cannot be reduced, then other features, which serve to smooth the force flow, should be added.

Use of these guidelines is illustrated in Fig. 5.6. Note that the improvements indicated can be qualitatively verified by sketching lines of force flow. It is also interesting to note that in all of these cases, the undesirable geometry/force flow interaction has been reduced and the component strengthened by removing material. This example illustrates the following corollary to the design principles:

> *Design load carrying components so that, in regions of high stress, necessary change in the direction of force flow is smooth and gradual.*

5.7 FORCE FLOW/STIFFNESS INTERACTIONS

Components deform under load depending on their stiffness. In many product assemblies and built-up structures such as weldments, spot welded structures, and riveted structures, redundant load paths may exist. A redundant load path is one that could be removed and still leave the product or structure in static equilibrium. Redundant load paths can lead to undesirable interactions because *load divides in proportion to the stiffness of the load path.*

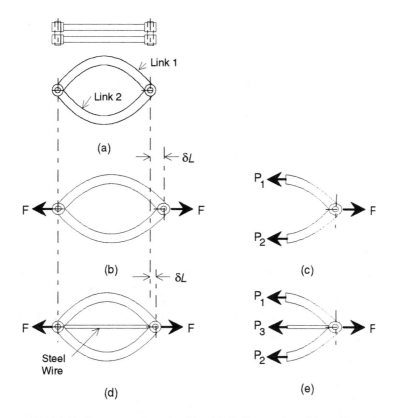

Figure 5.7 (a) Built-up tensile assembly, (b) built-up assembly subjected to an external force, (c) free-body diagram for two-link assembly, (d) built-up assembly reinforced with steel wire, (d) free-body diagram for three-link assembly.

This fact can be illustrated by considering a simple built-up assembly (Fig. 5.7a). Suppose the assembly is loaded by an external tensile load causing the assembly to elongate by an amount δL (Fig. 5.7b). The load carried by each bar is found from a summation of forces acting on the free-body (Fig. 5.7c),

$$F = P_1 + P_2 \tag{5.1}$$

Because the links are pinned together, the change in length of each link due to the external force must be the same. Remembering that stiffness (spring rate) is equal to the change in force divided by the change in length gives,

$$\frac{P_1}{k_1} = \frac{P_2}{k_2} \tag{5.2}$$

where k_1 and k_2 are the stiffness of links 1 and 2, respectively. Solving Eqs. (5.1) and (5.2) simultaneously gives,

$$P_1 = \left(\frac{k_1}{k_1 + k_2}\right) F \tag{5.3}$$

and

$$P_2 = \left(\frac{k_2}{k_1 + k_2}\right) F \tag{5.4}$$

Suppose that $k_1 = k$ and $k_2 = 2k$. Substituting these values into Eqs. (5.3) and (5.4), we find that $P_1 = 0.33F$ and $P_2 = 0.67F$. This shows that the load division is proportional to stiffness as expected. In general, for n redundant load paths, the amount of external load (F) carried by each load path depends on the stiffness of the load path and is given as,

$$P_j = \left(\frac{k_j}{\displaystyle\sum_{i=1}^{n} k_i}\right) F \tag{5.5}$$

5.7.1 Strength/Stiffness Relationship

Suppose now that in order to reduce the change in length δL due to the external load, a third link is added to the assembly in the form of a steel wire, which has stiffness $k_3 = 7k$ (Fig. 5.6d). Using Eq. (5.5), we find that the load carried by each link is now $P_1 = 0.1F$, $P_2 = 0.2F$, and $P_3 = 0.7F$. Hence, this third link, which is much stiffer, now carries the majority of the external load. An undesirable interaction arises in this situation if the wire is not strong enough to carry this large share of the load.

Juvinal (1983) illustrates this undesirable interaction with the following example. Suppose that when installed as part of a machine or structure, an angle iron has inadequate rigidity in that the $90°$ angle deflects more than

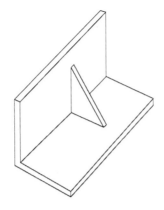

Figure 5.8 Web reinforcement added to an angle iron (Juvinal, 1983).

desired, although this does not cause breakage. To reduce the angular deflection, a small triangular web is welded in place (Fig. 5.8). In this redesign, the web is a redundant support that limits angular deflection. But it may well add stiffness far out of proportion to its strength. Cracks may appear in or near the welded joint, thereby eliminating the added stiffness. Furthermore, the cracks so formed may propagate through the main angle iron. If so, the addition of the triangular "reinforcement" would actually weaken the part.

Juvinal goes on to observe that failures in a complicated structural member (such as a casting) can sometimes be corrected by removing stiff but weak portions (such as thin webs). This is provided the remaining portions of the part are sufficiently strong to carry the increased load imposed upon them due to the increased deflection, and that the increased deflection that results is acceptable.

Based on these examples, we can state the following corollary to the design principles:

Design redundant load carrying members so that the strength of each member is approximately proportional to its stiffness.

5.7.2 Residual Stress in Built-Up Assemblies

Assuming that links in a pinned assembly are made using a mass production process, it is likely that the center distance between holes on each link will be slightly different (Fig. 5.9). To be assembled, therefore, one or both links must be elastically deformed to align the holes and allow the connecting pins to be

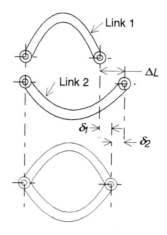

Figure 5.9 Deformations required to assemble two links of unequal length (note that the deformations are greatly exaggerated).

inserted as shown. After assembly, the center distance between holes on each link will be equal and such that the internal force generated in each link by the assembly process are in equilibrium. Because the center distance after assembly is different from the original center distances for the separate links, we see that a residual internal force or stress has been locked into the assembly as a result of the assembly process. The magnitude of this locked-in internal force depends on the stiffness of the load paths and is determined using an analysis similar to that given above to be,

$$P = \frac{\Delta L}{\dfrac{1}{k_1} + \dfrac{1}{k_2}} \tag{5.6}$$

where ΔL is the difference in hole center distances between the two links as shown in Fig. 5.9 and k_1 and k_2 are the stiffness of each link.

In general, residual stress should be avoided or held to a minimum when possible. It can weaken the structure when it adds to stress induced by external loads. It increases information content by causing distortion and other quality problems and by creating uncertainty and randomness. In some extreme cases, it can result in component failure during manufacture. For example, in a built-up assembly (Fig. 5.9), the link having the larger center distance will be loaded in compression when assembled and may buckle if deformed to severely.

An interesting insight regarding the effect of load path stiffness on the magnitude of the residual stress can be gained as follows. Let $k_1 = k$ and $k_2 = nk$. Substitution into Eq. (5.6) gives,

$$P = \frac{k\,\Delta L}{\left(1+\dfrac{1}{n}\right)} \tag{5.7}$$

Eq. (5.7) shows that the magnitude of the internal force locked in by the assembly process depends on the relative stiffness of the structural members. Specifically, by making one link or load path relatively stiff compared to the others, the magnitude of the locked-in force approaches the force required to deform the most flexible (least stiff) link. This implies following:

In built-up structures having redundant load paths, the undesirable effects of manufacturing process induced residual stress can be reduced by making one load path relatively stiff compared to all the others.

5.7.3 Precision of Built-Up Assemblies

Returning once again to the built-up assembly of Fig. 5.9, we see that the final dimensions of the assembly will depend on the relative deformation of each of the links. The deformation of links 1 and 2 are $\delta_1 = P/k_1$ and $\delta_2 = -P/k_2$, respectively, where the negative sign indicates that link 2 has the larger of the two original center distances. By again letting $k_1 = k$ and $k_2 = nk$, and using Eq. (5.7) to calculate the internal force P, the deformations of links 1 and 2 become,

$$\delta_1 = \frac{\Delta L}{1+\dfrac{1}{n}} \tag{5.8}$$

and

$$\delta_2 = \frac{\Delta L}{1+n} \tag{5.9}$$

Eqs (5.8) and (5.9) show that as link 2 is made very stiff compared to link 1, that is, as n is made large, then $\delta_1 \to \Delta L$ and $\delta_1 \to 0$. In other words, when one link is very rigid and the other very flexible, the flexible load path absorbs all or most of the deformation necessary to accommodate dimensional mismatch during assembly.

This insight has important ramifications for built-up structures such as automobiles and refrigerators, which are composed of a flexible skin and a rigid underlying frame. Because the rigid frame deflects very little upon assembly, all or most of the dimensional mismatch between the frame and the skin is absorbed by the skin, which is flexible and therefore deforms easily. As a consequence, the rigid frame determines the final dimensions and/or shape of the assembly. Therefore, to insure dimensional fidelity and high quality of the final assembly, the rigid member should be made very accurately since it establishes the final dimensions. This leads to the following corollary:

Design redundant load carrying members so that the most rigid members are also the most precise (i.e., dimensionally accurate).

5.7.4 Stress Concentration Due to Stiffness Mismatch

Another undesirable interaction that can occur due to stiffness of redundant load paths is severe stress concentration. This type of undesirable interaction generally occurs in situations where load is being transferred from one component to another and the deformation of the related components is poorly matched. Consider, for example, a shaft and hub assembly (Fig. 5.10). Stress concentration will be high if the hub is rigid and deflects little at *A* where the shaft deflection relative to the hub is greatest (Fig. 5.10a). By making the hub stiffness better match the shaft stiffness, load is transferred more gradually and smoothly (Fig. 5.10b). Poorly matched deformations deflect the force flow while matched deformations smooth it. This insight is summarized by the following corollary (Pahl and Beitz, 1984):

Design related components in such a way that, under load, they will deform in the same sense and, if possible, by the same amount.

5.8 DESIGN FOR MANUFACTURE AND ASSEMBLY

The design principles can be used to both explain and provide insight into design for manufacture and assembly practices and guidelines. According to Boothroyd, Dewhurst, and Knight (1994), reducing part count and part type is the most important design for assembly guideline. This is no surprise when it is realized that reducing or minimizing the number of individual parts and part types in a product is perhaps the most effective way to reduce information content of the product. A component that is eliminated costs nothing to design,

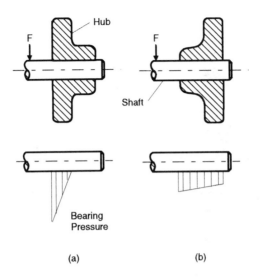

Figure 5.10 Force transmission from shaft to hub: (a) severe stress concentration due to poorly matched deformations; (b) more even bearing pressure because of matched deformations.

make, assemble, move, handle, orient, store, purchase, clean, inspect, rework, or service. It never jams or interferes with automation. It never fails, malfunctions, or needs adjustment. It requires no drawing or part number and never needs to be changed or redesigned.

When viewed in this way, it is seen that many of the widely advocated design for manufacture and assembly guidelines (see Chapter 14) are really ways of reducing manufacturing information content. For example, reduced information content explains the emphasis on near net-shape manufacturing processes and the importance of avoiding or minimizing secondary processing. Often, higher material and/or processing cost can be accepted when secondary processing is avoided because the reduced information content that is realized leads to lower overall production cost and systems cost.

In automation applications, separate fasteners are difficult to feed, tend to jam, require monitoring for presence and torque, and require costly fixturing, parts feeders, and extra stations. Avoiding separate fasteners eliminates all this extra information content. The same is true for assembly directions and material handling. By designing for top-down, Z-axis insertion, many needed degrees of freedom and the information that accompanies them are eliminated. Similarly, the use of symmetry to assist in achieving proper part orientation and the use of parts magazines, tube feeders, part strips, as well as palletized trays and kitting techniques for preserving orientation greatly reduces the information required to manipulate and handle parts. Finally, use of generous tapers and other guiding, orienting, and locating features in part design greatly simplifies component insertion and assembly because it reduces the amount of information needed to perform these tasks.

5.9 KEY TAKEAWAYS

By avoiding undesirable interactions and driving toward simple design solutions, the design team ensures good design.

- The undesirable interaction principle and the minimum information principle are underlying fundamental principles of good design. Products that are designed following these principles and are in harmony with them will invariably have higher total product quality and lower total cost and time than product designs that violate them.
- In developing a good design, each design decision should be guided by the design principles and must not violate them. One way to do this is to first analyze a proposed design solution for undesirable interactions and then develop redesigns that avoid all identified undesirable interactions. The simplest redesign solution is the preferred or most acceptable design solution because it has the least information content.
- There are a variety of corollaries to the design principles. These corollaries provide additional design guidance that can help to quickly identify the best design approach in particular design situations.
- The design principles and corollaries explicitly state what is intuitively obvious to experienced designers and commonly used by good designers. Such a system of principles, therefore, provides a firm basis on which design knowledge can be expanded in a systematic way.

6

Total Cost Reduction

6.1 INTRODUCTION

Total cost is the sum of all costs, both direct and indirect, that result from the design, manufacture, distribution, sale, service, use, and disposal of the product over its life cycle. It is well known that a large percentage of total cost is determined by the product design, especially early design decisions that establish the particular physical concept and part decomposition to be employed. Unfortunately, it is often difficult or impossible to evaluate the impact of these early design decisions on total cost. This is because these decisions effect indirect cost much more than direct cost. Indirect cost is usually very difficult or impossible to estimate in the early stages of design, when most far-reaching design decisions are made.

So how can the manufacturing enterprise make quality early design decisions that reduce total cost when the effect of early design decisions on total cost can't be estimated? In this chapter, we seek to answer this question by using information content as a surrogate for total cost. To do this, we first discuss the shortcomings of current cost estimation methods. We then show that information content is equivalent to total cost. This allows us to infer that design decisions that reduce information content also reduce total cost. These results are then used to develop a total cost reduction strategy based on reduction of information content. We call this strategy "guided common sense."

6.2 COST ESTIMATION

The ability to estimate cost as a function of design decisions is essential for reducing total cost by design. Unfortunately, estimating cost is an inexact science, especially in the early stages of design when design information is very tentative and incomplete. In addition, most of the cost estimation techniques available are based on historical data and often do not include indirect cost or consideration of cost consequences that may arise as a result of quality or manufacturing problems. Never the less, it is common practice to estimate cost and to use these costs as a basis for making many far-reaching design decisions.

6.2.1 Traditional Cost Estimation Methods

Traditional cost estimation methods typically used in the early stages of design include parametric methods and empirical methods. In parametric cost estimation, the overall cost of a product is correlated with major product features such as weight. For example, the overall cost of a new airplane can be estimated by using a dollar per pound figure derived from production costs of similar aircraft. In empirical cost estimation, the cost of a component or assembly is estimated based on the cost of specific design features which have been determined using statistical correlation of the cost of past designs with these features. Empirical methods are very useful for comparing costs of alternative designs and are commonly used for that purpose. See for example, Boothroyd, Dewhurst, and Knight (1994), Boothroyd (1992), Dixon and Poli (1995), and Ostwald (1988).

Unfortunately, there are deficiencies inherent in both the parametric and empirical cost estimation methods. Parametric cost estimation methods are only valid for narrowly defined domains and they provide little or no guidance to the design team regarding the acceptability of detail design decisions. Empirical methods are useful in providing design guidance when performing detail design of individual parts. They are generally not very helpful, however, in guiding early part decomposition decisions or deciding initial configuration design. Both methods are also limited in scope in that they tend to focus on direct cost rather than total cost. As a result, these methods provide little insight or guidance regarding systems cost and other indirect cost.

6.2.2 Activity Based Cost Analysis

Activity based cost (ABC) analysis is an alternative cost system used by management in some companies to associate indirect cost with particular products and components (see Fig. 6.1). In the ABC method, overhead dollars spent by the various activities within the manufacturing enterprise are apportioned to specific products and components that are manufactured by the company. For example, if X amount of dollars is spent on inspection activities, the ABC method will help determine how much of this money is spent on each product or component made. This allows management to determine which products are truly profitable and which are costing more in overhead dollars than they generate in revenue. Most importantly, it allows management to

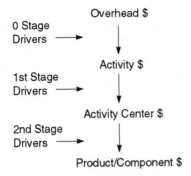

Figure 6.1 In an activity based cost (ABC) system, overhead expenses are attributed to individual products and components by identifying the hierarchy of cost drivers that cause indirect dollars to be spent and then determining the indirect dollar amounts that are actually spent on each product manufactured.

better manage the overall business as well as the overhead activities associated with each product's manufacture.

6.2.3 Design Decisions and Indirect Cost

With the right ABC system, there is the potential of providing the design team with the knowledge they need to assess the impact of design decisions on indirect cost. Unfortunately, creation of ABC systems to do this has not been a high priority and so this potential is largely unfulfilled at present. Hence, the real problem we are faced with is an inability to estimate, even on a very rough basis, the impact of early design decisions on indirect cost.

This inability has resulted in a variety of undesirable behaviors. Consider, for example, the price structure of a modern manufacturing company (Fig. 6.2). The selling price can be divided into the direct cost, indirect cost, tax, and profit. Often, a business is able to establish nominal ratios between the material cost, the direct cost, and the selling price based on an analysis of current costs. A typical ratio might be 1:3:9. This implies that a product containing $1 of raw material costs $3 to make, and has to be sold for $9 in order to make a reasonable profit. Because of the cascading effect implicit in such ratios, a small reduction in material cost or direct cost could lead to a large reduction in selling price, or if selling price is constant, a large increase in profit. This, and the focus of most cost estimation methods on direct cost, has caused many design offices to over emphasize piece-part cost while accepting indirect cost as an unavoidable cost of doing business which is beyond their control.

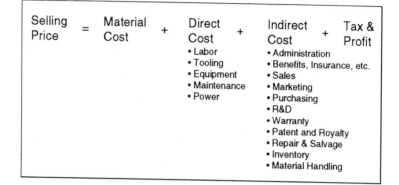

Figure 6.2 Price structure of a typical manufacturing enterprise. Commonly used figures of merit include the processing overhead ratio (direct cost/material cost) and the operational overhead ratio (selling price/direct cost). For many firms, these ratios have numerical values of about 3. Hence, to make a reasonable profit, selling price needs to be approximately 9 times material cost (Chow, 1978).

Our contention is that *early product design decisions drive both direct and indirect cost*. This means that we must not only consider the impact of design decisions on direct cost, but on indirect cost as well. Examples of indirect cost directly effected by design decisions include the cost of managing supplier relations, the cost to stock a wide range of raw material sizes, the cost to maintain and operate a hard to control manufacturing process, the cost to repair defective product, and the cost in loss sales due to unreliability, to list just a few. The total of these indirect costs can be much larger than direct cost as implied by the cost ratios discussed above. In addition, most indirect cost continues for the life of the product and is independent of the volume of product manufactured. Clearly, these costs cannot be ignored, especially if it is possible to reduce them by improving the quality of early design decisions.

Why does consideration of indirect cost present such a problem? Suppose a product currently composed of 1,000 unique parts is redesigned to provide essentially the same function using 100 unique parts, what will be the effect on total cost? We can calculate the direct cost savings by using empirical cost estimating methods to estimate the cost of the 100 unique parts and then comparing this cost with the direct cost of the current product. Even though the part count reduction is very large in this example, it may be found that direct cost is not reduced significantly, or that it may have even increased.

To illustrate, suppose that one way parts are eliminated in this new design is by replacing a built-up structure consisting of several stamped sheet metal parts riveted together with a plastic injection molding. In the built-up structure, each part may be very inexpensive. In addition, production machinery such as stamping presses and riveting machines are in place and the expertise to operate them is available. Use of a plastic injection molding, on the other hand, may add considerable material and tooling cost compared to the stamped metal parts and the part may need to be purchased from an outside supplier. As a result, analysis, using conventional cost estimation techniques, may show that the direct cost has increased rather than decreased. Based on the cost ratios mentioned above, an increase in material and direct cost would likely prove unacceptable in many design organizations.

How much has been saved in indirect cost? Suddenly, we have a problem because there are no cost estimation methods available to help. Using common sense, however, we reason that indirect cost must be significantly reduced since all the costs associated with designing, making (or purchasing), stocking, handling, assembling, and repairing 900 parts (1,000 − 100) as well as the tooling costs including maintenance and repair and the quality risks associated with their assembly has been eliminated. In fact, the cost saved in reduced material handling alone may more than offset the increased material cost of the plastic injection molding discussed above.

These examples illustrate the confusion and conflicts that can arise due to the current inability of most design organizations to measure the cost impact of design decisions on total cost. To avoid these difficulties, an entirely different approach is needed, one that focuses on total cost rather than direct and indirect cost.

6.3 INFORMATION CONTENT AND TOTAL COST

What can be used in place of cost to help guide design decisions? The answer to this question is implied by the principles of good design discussed in Chapter 5. A product designed using these principles will be inherently easier to design, manufacture, maintain, service, and support over its life cycle. Such a product will naturally have the lowest possible total cost. In other words, total cost is reduced by reducing the information content of the design, provided that all undesirable interactions are also avoided.

6.3.1 Equivalency of Cost and Information Content

As a starting point, we must decide how to measure the information content of a given part or assembly. One idea, due to Wilson (1980), is to express information content (I) in terms of dimensions and tolerances. From information theory, the information content of a design is calculated as

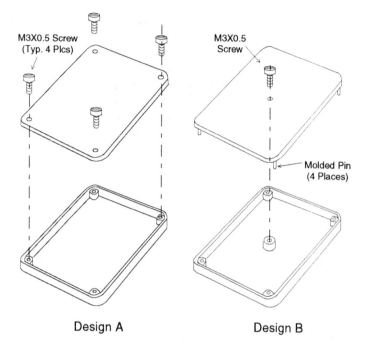

Design A   Design B

Figure 6.3 Alternative designs for securing a cover to an enclosure using threaded fasteners.

$$I = \sum_{i=1}^{n} log_2 \left(\frac{d_i}{t_i} \right) \qquad (6.1)$$

where d_i is a particular dimension, t_i the tolerance on that dimension, and n the total number of dimensions required to fully specify the part. For assemblies, the dimensions and tolerances refer to the placement of each part in the assembly. Since information is traditionally measured in bits, the log is taken to the base 2.

To illustrate how information content can be calculated using Eq (6.1), consider Design A shown in Fig. 6.3. In this design, four M3×0.5 screws secure the cover of a square box. Suppose the box is 35±0.5 mm long on each side and that the mounting holes are located on 30±0.5 mm centers. Suppose also that the tolerance on the M3×0.5 screw is ±0.063 mm and that the screws are tightened to a tolerance of 60°/360° = 1/6th of a turn. The dimensional information content of the four screw diameters is calculated as:

$$I_{Screw\,Diameter} = 4\log_2\left(\frac{3}{0.063}\right) = 4\,(5.5735) = 22.294 \text{ bits}$$

Performing similar calculations for each assembly dimension and summing gives:

Information Content of Design A in Fig. 6.3

Feature	Quantity	Dimension	Tolerance	Information (bits)
Screw Diameter	4	3 mm	0.063 mm	22.294
Position screw in X-direction	4	30 mm	0.5 mm	23.628
Position screw in Y-direction	4	30 mm	0.5 mm	23.628
Install screw (6 turns)	4	2160°	60°	20.680
Locate cover in X-direction	1	35 mm	0.5 mm	6.129
Locate cover in Y-direction	1	35 mm	0.5 mm	6.129
Total				102.488

In an alternative design, shown as Design B in Fig. 6.3, the cover is attached using a one-screw assembly with molded pins to locate the cover on the box. Information content of this design is calculated as:

Information Content of Design B in Fig. 6.3

Feature	Quantity	Dimension	Tolerance	Information (bits)
Screw Diameter	1	3 mm	0.063 mm	5.573
Position screw in X-direction	1	17.5 mm	0.5 mm	5.129
Position screw in Y-direction	1	17.5 mm	0.5 mm	5.129
Install screw (6 turns)	1	2160°	60°	5.170
Locate cover in X-direction	1	35 mm	0.5 mm	6.129
Locate cover in Y-direction	1	35 mm	0.5 mm	6.129
Total				33.259

Another design alternative would be to use snap-fits to secure the cover to the box. Such a design integrates the fastening and locating functions into the cover and box and eliminates the need for separate fasteners. The information content of the snap-fit assembly is calculated to be:

Information Content of Snap-Fit Design

Feature	Quantity	Dimension	Tolerance	Information (bits)
Locate cover in X-direction	1	35 mm	0.5 mm	6.129
Locate cover in Y-direction	1	35 mm	0.5 mm	6.129
Total				12.258

The time to manually assemble these designs can be estimated using the Boothroyd-Dewhurst Design for Assembly Method (Boothroyd, Dewhurst, and Knight, 1994). In this method, the time required to handle and insert each part is estimated by analyzing the part geometry. These times are then summed to calculate the assembly time for the design. The results for the three designs are summarized as follows:

Assembly Time for Design A in Fig. 6.3

Part	Quantity	Handling Time (sec)	Insertion Time (sec)	Assembly Time (sec)	Comments
Base	1	1.95	1.50	3.45	
Cover	1	2.36	6.50	10.68	Not easy to align
Screws	4	1.80	8.00	39.20	
Total				53.33	

Assembly Time for Design B in Fig. 6.3

Part	Quantity	Handling Time (sec)	Insertion Time (sec)	Assembly Time (sec)	Comments
Base	1	1.95	1.50	3.45	
Cover	1	2.36	2.00	4.36	Easy to align
Screws	1	1.80	8.00	9.80	
Total				17.61	

Assembly Time for Snap-Fit Design

Part	Quantity	Handling Time (sec)	Insertion Time (sec)	Assembly Time (sec)	Comments
Base	1	1.95	1.50	3.45	
Cover	1	2.36	2.00	4.36	Easy to align
Total				7.81	

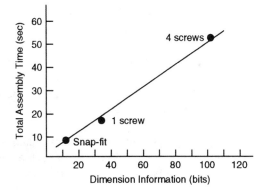

Figure 6.4 Plot of total estimated assembly time verses dimensional information content for three alternative cover attachement designs.

Hoult and Meador (1996) have shown that, if parts are made by a single manufacturing process, the average time (T) to fabricate the part is

$$T = A I \tag{6.2}$$

The same result also applies for assemblies made with a single assembly process such as manual assembly. A plot of assembly time verses information content verifies this for the cover assembly example (Fig. 6.4). In general, the coefficient (A) in Eq. (6.2) must be determined from past cost data or from an analysis such as that conducted above. For the cover assembly example,

$$A = \frac{53.33 - 7.81}{102.488 - 12.258} = 0.5045 \text{ sec/bit}$$

Since processing time and processing cost are directly related, these results clearly indicate a direct relationship between information content and cost. Most importantly, they demonstrate that cost is reduced as information content is reduced.

6.3.2 Strategies for Reducing Information Content

The amount of information content in a given part, or subassembly, or product, or manufacturing system, is determined by design decisions. Equation (6.1) tells us that if we eliminate dimensions, relax tolerances, or both, dimensional information content will be reduced. And, because information content is equivalent to cost, this will in turn, decrease total cost. The concept of information content therefore provides the needed connection between design decisions and total cost.

Dimensions and tolerances, are just one source or measure of information content. In most practical situations, there appears to be an almost limitless number of different sources. For products composed of discrete parts, for example, some of the more important sources of information content include:

- The number of separate operations or activities, number of instructions per activity, and the number of repetitions of each activity required to manufacture and assemble a particular product, subassembly, or component.
- The number of different tools and processes utilized in a product or component manufacture.
- The number of unique features, facets, characteristics, functional surfaces, etc. contained in a component.
- The number of interfaces and interactions between assembled components.
- The amount of part-to-part variability and product-to-product variability.
- The amount of randomness or variability associated with manufacturing processes, inspection processes, testing methods, material handling, order processing, shipping and warehousing, and all other activities associated with product manufacture.
- The amount of randomness or variability associated with product operation, servicing, maintenance, and disposal.

The seemingly endless sources of information content make it appear that calculating information content is no easier or more straightforward than calculating total cost. In our view, however, the ability to calculate information content is not what is important. Rather, the key is to understand how information content is effected by design decisions and to know how to reduce information content by making the right design decisions.

In looking at the sources of information content listed above, it is not hard to imagine an equation similar to Eq. (6.1) that would express the total amount of information contained in each different source. Suppose, for example, that there are (m) different sources of information content and that there are (n_j) individual components contributing to each source. Suppose further that the amount of information contained in each individual component is measured by a quantity (α_{ij}) where each (α_{ij}) is an appropriate measure for the type of information contained in a particular source. Then, total information content would be computed as

$$I_{Total} = \sum_{j=1}^{m} \left(\sum_{i=1}^{n_j} \log_2 \alpha_{ij} \right)_j \qquad (6.3)$$

Equation (6.3) is very explicit regarding what must be done to reduce information content: *eliminate and simplify*. Each (α_{ij}) of a product component or assembly is a *source component* of information content. The numerical value of $(\log_2 \alpha_{ij})$ represents the *amount* of information contained in each individual source component. Total information content is the linear sum of the information contained in each source component summed over the number of different sources. Total information content can therefore be reduced by decreasing the number of components (n_j) that contribute to each source, by decreasing the amount of information $(\log_2 \alpha_{ij})$ contained in each component, and, when possible, by decreasing the number of different information sources (m) that contribute to the overall information content of the design.

Therefore, information content, and total cost, of a product design is reduced by doing the following: (1) *eliminate* sources of information content by reducing (n_j) and (m), and (2) *simplify* the design by decreasing the amount of information $(\log_2 \alpha_{ij})$ contained in each source component. These observations can be generalized as the following fundamental strategies for information content reduction:

1. *Eliminate* sources of information content.
2. *Simplify* by reducing the information contained in the sources that remain.

The validity of these strategies can be confirmed by comparing Design A and Design B shown in Fig. 6.3. Dimensional information content is reduced from about 102 bits for Design A to about 33 bits for Design B by eliminating 3 screws, centering the one remaining screw to simplify screw location, and using molded pins to simplify alignment of the cover. It is also evident from the analysis that manually assembled screws contain a considerable amount of information. By eliminating separate fasteners altogether, assembly information content is reduced even further to about 12 bits for the snap-fit design.

Limiting the variety of each source of information content by *standardizing* them also reduces information content. Standardization reduces information content by reducing numerous options to a manageable few. However, if not carefully implemented, standardization can increase information content by creating conflicts and limiting alternatives. Standardization must therefore be implemented with great care and forethought. With this caveat in mind, a third fundamental strategy for information content reduction is:

3. *Standardize where possible* to limit sources of information content and the amount of information contained in each source.

The three fundamental guidelines, *eliminate, simplify,* and *standardize where possible* are by no means new ideas or concepts. They have been widely practiced for many years and underlie most of the cost reduction methods and strategies employed in value engineering, producibility engineering, and design for manufacture. By showing that these strategies work by reducing information content, however, we have established a clear and irrefutable connection between these best practices and total cost reduction. We can therefore apply the strategies in any design situation with great confidence that they will, in all cases, reduce total cost. In addition, the meaning of each term is now explicit and unambiguous: to *eliminate* means to eliminate sources of information content; to *simplify* means to reduce the amount of information contained in an information source; to *standardize* means to limit the variety of sources.

To reduce total cost, these strategies should be applied to all aspects of the manufacturing system. This includes the product, the manufacturing equipment, processes, material handling, and so forth used to produce it, and the material flows, information flows, and energy flows that transform starting materials into finished product.

6.4 GUIDED COMMON SENSE

Applying the simple information reduction strategies is largely a matter of common sense. Because the "eliminate," "simplify," and "standardize where possible" strategies are based on the underlying concept of information content reduction, we call this approach *guided common sense*. In essence, we reduce total cost by using common sense guided by the principles of good design discussed in Chapter 5. A product designed to minimize information content, while also avoiding undesirable interactions, will be inherently easier to design, manufacture, maintain, service, and support over its life cycle. Such a product will naturally result in the lowest possible total cost.

Many designers may recognize this approach as the "KISS (keep it simple stupid)" principle of design. The problem with KISS is that it hard to consistently apply because there is no organized system of knowledge and best practices to guide it. Guided common sense implements the KISS principle by formulating clearly articulated design strategies that are based on the principles of good design. Guided common sense is, in essence, the underlying key to achieving low total cost. We apply in it a variety of ways throughout the remainder of this book.

6.5 KEY TAKEAWAYS

Total cost is reduced by reducing the information content of the design.

- Total cost is directly related to information content.
- Total cost decreases as information content is reduced.
- Information content is reduced by (1) eliminating sources of information content, (2) reducing the information content in the sources that remain, and (3) standardizing where possible to further limit sources of information content and the amount of information contained in each source.
- Common sense can be reliably used to judge the effect of alternative design decisions on information content. Therefore, an ability to accurately compute information content is not essential.
- The guided common sense approach consists of three strategies: "eliminate," "simplify," and "standardize where possible."

7

Design Process Improvement

7.1 INTRODUCTION

An effective product development process has the following characteristics:

1. The products developed are "winners" in the marketplace. Winning products are highly desirable products that meet well defined and clearly understood customer needs and are clearly superior to best in class competitive products.
2. The products are developed and introduced in a timely manner so that the full market share and profit potential is harvested.
3. During the development process, there are no "surprises." The product transitions easily and smoothly into production with minimal engineering change. All performance and quality goals are achieved.
4. The product yields a reasonable profit over its life cycle.

In this chapter, we explore the underlying nature of the product development process to understand how it can be improved. We then use this understanding to recommend product development practices that will help ensure an effective and efficient product development process.

7.2 DESIGN ITERATION

Product design is a process. It begins with the recognition of a need and ends with a manufactured product that is accepted by customers as being satisfactory in service. The process proceeds from the abstract to the concrete. This progression can be divided into a series of phases in which information about the design is developed and modified (Fig. 7.1).

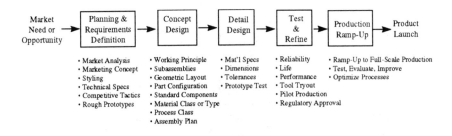

Figure 7.1 Phases of the product realizaiton process.

The design process is also a learning process that involves making design decisions, evaluating their acceptability, and then making modifications where necessary. This design-analyze-redesign procedure is an inherent characteristic of the design process that is often referred to as the "iterative nature of design." In general, at least three iteration cycles may be required (Fig. 7.2). The first cycle of planning, design, manufacture, and assembly is undertaken to produce a model, which can be market and performance tested. Results of this initial investigation may indicate ways of improving the design. This information is incorporated in a new cycle either to manufacture an improved model for further tests (dashed line), or to optimize and refine the engineering design and product reliability in a second major cycle. It may then be worthwhile to further refine the design in a third major cycle to optimize the manufacturability and serviceability of the product, before finally putting the finished design into full-scale production.

Design iteration is an essential learning process. When design iteration occurs late in the process, however, it can also be a source of delays and unplanned for costs if not managed properly. Managing iteration is therefore the key to effective and timely product development.

7.3 ENGINEERING CHANGE AND THE RIPPLE EFFECT

Design iteration can result in local design changes. These local changes generally propagate in a *ripple effect* throughout the design because they require that each part of the design affected by the change be reexamined. The unavoidable consequence is *engineering change* caused both by the design iteration itself and the ripple effect it produces. In the early phases of a design project, engineering changes are handled fairly easily because the design is fluid, hardware is still remote, few people are involved, and constraints and interactions have not become tight. In later project stages, engineering changes become much more difficult to handle because so much has been designed and irrevocably fixed. Engineering changes made late in a design project can have

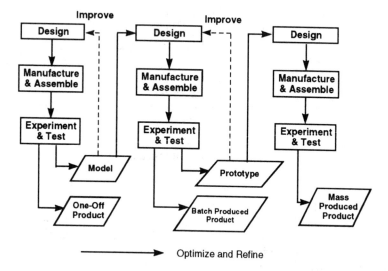

Figure 7.2 Cycles in product creation (VDI Guideline 2221, 1987). The key to effective and efficient product development is management of these iterative product development cycles.

extremely disastrous consequences. Because of the ripple effect, an ill-chosen design solution for even a relatively minor design problem can put the whole project in jeopardy. For these reasons, the range of possible solutions to design problems discovered late in the project are severely limited and almost any design solution chosen will result in undesirable compromises and suboptimal design. This can, in turn, result in increased cost and customer dissatisfaction that continues for the life of the product.

The cost of making engineering change late in the project can also be enormous. Late changes are more costly because time delays are more significant, more personnel and possibly suppliers are involved, and more parts and systems are involved because of the ripple effect. When engineering changes occur after production release or originate as a result of production problems, the costs incurred include additional indirect costs such as scrap parts and idle machinery and workforce. Engineering change made after sales can be the most costly of all because of high warranty or service costs and invisible costs such as loss of reputation and declining workforce morale.

The cost saving potential of avoiding design change late in the project can be dramatic (Fig. 7.3). Assume that the cost of making an engineering change in the conceptual phase is $1.00 and that it increases by a factor of 10 with each subsequent phase. Assume also that a total of 100 engineering changes are

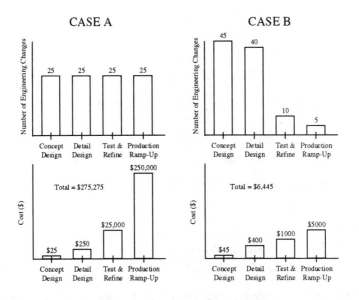

Figure 7.3 Cost saving potential of performing engineering change early in the design process. Cost is estimated using the law of 10's. (Not to scale.)

processed over the duration of the project. If the changes are uniformly distributed over each phase of the project, cost of engineering change will be $275,275 (Case A, Fig. 7.3). If, on the other hand, most of the changes are made in the early stages of design, cost for the same 100 engineering changes will only be $6,445 (Case B, Fig. 7.3). Design time is also greatly reduced because early changes are easier and less time consuming to perform (Fig. 7.4).

7.4 PRESCRIPTION FOR IMPROVEMENT

Engineering change is usually the unfortunate consequence of uninformed or poor quality design decisions. A design decision is uninformed when it is made in the absence of necessary information or when important aspects of the decision are nor properly considered. A part design that must be changed because it is difficult or impossible to manufacture is a prime example. Had the constraints of the manufacturing process been properly considered when the part was originally designed, subsequent design changes to improve manufacturability would not be required. Avoiding uninformed design decisions and the engineering change that results is the primary goal of design process improvement.

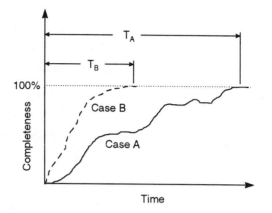

Figure 7.4 Design cycle time reduction potential of performing engineering change earlier in the design process.

A well-managed design process avoids the need to make engineering change late in the project by facilitating informed design decision making. A design decision is informed when all necessary information is available and all aspects of the decision are systematically considered. Experience has shown that informed decision making requires two key ingredients: (1) good communication and (2) design discipline. Good communication ensures the availability of needed information and design discipline ensures that all relevant information is systemically considered. As a general prescription for improving the product design process, we recommend that communication and design discipline be improved by implementing the following practices.

1. **Use a team approach.** The goal of the team approach is to enhance communication and ensure that all needed information is available when critical design decisions are made.
2. **Develop a review based development process.** Formal design reviews help ensure that the best possible design decisions are being made, that all aspects have been properly considered and that the project is ready to proceed to the next phase.
3. **Develop a design guideline.** The purpose of the design guideline is to ensure a disciplined and systematic approach that is common to all projects and understood by all team members. The guideline typically includes a design process roadmap, simple checklists, recommended design methodologies and tools, and needed information about "downstream" constraints and process requirements.

How and if these practices are implemented depends heavily on the particular manufacturing firm involved. *One size does not fit all.* Each firm started with a particular set of initial conditions that established the way product design is performed and then evolved over time under the influence of a variety of boundary conditions unique to its culture and business. We therefore cannot say with any degree of certainty that one practice or way of doing things is better than another. Each firm has to find its own path. There are, however, certain characteristic features or "success factors" associated with each of the above practices that appear to be desirable. We discuss these characteristics in the sections that follow.

7.5 TEAM APPROACH

The purpose of the team approach is to ensure that all needed information is readily available as design decisions are made throughout the course of the project. This means that the team should be cross-functional so that all points of view and disciplines are represented. Cross-functional teams enhance design creativity due to cross-fertilization of thought processes, behaviors, and functional skills. When functioning properly, the cross-functional team approach allows the development process to occur in nonlinear iterations that bounce back and forth between disciplines so that design decisions are fully informed (Fig. 7.5). Without an effective team approach, the design is likely to proceed in a more linear fashion. This leads to "surprises" that require extensive iteration and engineering change. The lengthy and often costly "over-the-wall" iterations that can occur between design and manufacturing is representative of this phenomenon.

As the project goes through its various phases (Fig. 7.1), different skills and numbers of people are required. Also, experience has shown that, in general, teams should be small. A small team has anywhere from 2 to 25 members, but less than 10 members is often best. To accommodate fluctuating resource needs and team size, the project team can be divided into a core team and an extended team. The *core team* is composed of a small number of people with complementary skills who have day-to-day responsibility for the project including budget and schedule control. Core team members are involved with the project from the start and typically continue until the product is tooled and in production. The *extended team* includes all those who bring needed skills to the project and are involved at particular stages of the project or who participate on an as needed basis. This includes component suppliers, manufacturing equipment developers such as automation systems houses, tooling suppliers, and others such as "lead users" and product service providers.

A serious project deserves that serious attention be paid to the selection of core team members. As a minimum, the core team will include representatives

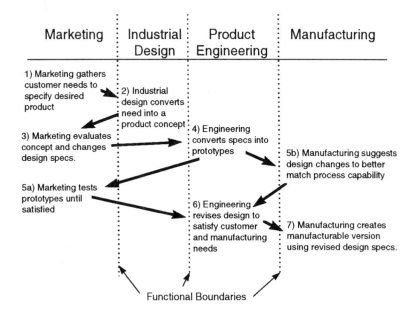

| Marketing | Industrial Design | Product Engineering | Manufacturing |

1) Marketing gathers customer needs to specify desired product

2) Industrial design converts need into a product concept

3) Marketing evaluates concept and changes design specs.

4) Engineering converts specs into prototypes

5a) Marketing tests prototypes until satisfied

5b) Manufacturing suggests design changes to better match process capability

6) Engineering revises design to satisfy customer and manufacturing needs

7) Manufacturing creates manufacturable version using revised design specs.

Functional Boundaries

Figure 7.5 Cross-functional design teams facilitate nonlinear interations instead of linear "over-the-wall" iterations.

from marketing, product engineering, and manufacturing. As a general guideline, the core team should include expertise from most areas of the business that play a role in or are effected by new product development. The project manager and other key team people should have full-time assignment to the project with no part-time diversionary duties. In matrix organizations where project staff are full-time reports to their functions and only dotted-line to the project, the project usually suffers unclear direction, conflicting orders, and missed schedules, as well as less than effective use of expensive talent (Ettlie and Stoll, 1990).

In addition to being cross-functional, successful product development teams typically have three additional qualities. First, all members of the team share a common purpose and set of goals that keep the team focused on results and set aspirations. Second, there is a common approach that every team member understands and follows. This facilitates working together, sharing information, and effective communication. Finally, all team members hold themselves mutually accountable. Trust and mutual accountability are crucial for good teamwork. Without these key ingredients, much of the product development advantages and leverage offered by the team approach is lost.

Another key factor in developing a successful team approach in many firms is the choice of project leader. James N. Hughes, a long time employee of General Electric, offers the following list of characteristics that are possessed by many successful team leaders (Ettlie and Stoll, 1990):

- Respected both for business judgement and technical expertise.
- Total dedication, to the point of obsession, with the particular project.
- Charisma that conveys a strong impression to others of knowing the direction to go, welcoming other travelers, and convincing them to follow.
- Stamina to continue with the project until all goals are met.
- Positioned to have authority to cause instant action without having to negotiate, arbitrate, and plead for support.

As stated previously, the main purpose of the team approach is to facilitate the communication required to perform nonlinear design iterations and make fully informed design decisions. Achieving effective communication both within the team and to supporting operations is however still critical. Just because the firm decides to form a project team and use the team approach does not guarantee that effective communication will occur. Experience has shown the following aids to be useful in helping to improve effective team communication (Ettlie and Stoll, 1990):

- Ensure that the core team stays together from inception to production. The team members know each other's views, contributions, and work status. Few people are added later so that less catching up is needed.
- Have team members sit together, work together, attack problems together, and devise solutions together. Geographical distance, whether within plants or between cities, is a barrier to effectiveness.
- Make joint plans, hold joint planning and status meetings, use common schedule formats.
- Drive toward maximum use of electronics for all interaction, including computer e-mail, networks, multi-access databases, progressive updating of data, and so forth. The present rapid advances in systems technology as applied to internal organizational communication are providing continual improvements.
- Push for physical demonstration of progress, not progress reports. Simplify communication: either it's there or it's not; either it works or it doesn't.
- Package work so that most of the communication is within the team. Then useful communication to others involves updates but not details.

7.6 FORMAL DESIGN REVIEWS

The purpose of conducting formal design reviews during the course of the design projects is to help ensure an optimal design. An *optimal design* is one that meets all customer requirements in an optimal way (high external quality) and also yields a satisfactory profit because it is easy to manufacture and support over its life cycle (high internal quality). Design reviews are formal in that they are scheduled by the team and are intended to be systematic evaluations of the design, not of schedule or budget performance. The design team (core team) must prepare for the review and knows "up front" what must be demonstrated in order to pass the review and proceed with the project.

A hard and fast rule that must be adhered to is that the project does not proceed to the next stage unless both the team and management agree that the design is ready. Three outcomes are possible:

- Agreement to proceed.
- Detailed plan, schedule, and resources to correct deficiencies.
- Project termination.

The goal obviously is to obtain agreement to proceed. However, if "show stopper" problems or concerns are identified, then a detailed plan to correct the deficiencies is developed and agreed upon, preferably as part of the review process. This plan outlines what needs to be done and includes a schedule and budget authorization. Most importantly, the plan specifies the acceptability criterion that must be satisfied as well as what should be done if the problem or problems can't be satisfactorily resolved. If it is clear from the design review that the project should not proceed, then a recommendation to terminate the project or redefine the project is also a viable outcome of the design review.

In general, it is recommended that knowledgeable people who are not directly involved in the project conduct the design review. This helps to ensure that the right questions are asked and that all ramifications and aspects are properly questioned and considered. The key concern for those conducting the review is to satisfy themselves and the firm that the project is on track. Questions that should be upper most in their minds include the following:

- Is the team following the process?
- Are the market and customer needs correctly understood?
- Is the proposed product concept responsive to customer needs and to issues of efficient manufacture?
- Are there latent defects? Omissions? Alternative approaches?

The reviewer's focus should be on thinking critically about the design. They should be tactful, but should not hold back. Asking the right questions about "why" and "how" can be crucial to the success of the project. The goal of the review is to not only surface potential problems, but to also impose design discipline on the team. If the team knows from experience that hard questions will be asked and that they will need to justify their design decisions, then they are more likely to make more informed decisions that are easily defended. The end result will be a better product design.

In presenting their design for review, the team should strive to explain the design to the fullest extent possible and to obtain as much feedback as possible. Some guidelines for presenting the design include:

- Explain the design
    - Provide supporting evidence
    - Discuss methods used
    - Discuss pros and cons of alternatives considered
    - Discuss trade-offs
- Explain problems and issues
- Seek creative input and advice
- Make sure all needed information is available

To ensure a successful review, the team must have an attitude that is open to criticism and receptive to suggestions. It is extremely important for the team to remember that the goal of the review is to ensure a winning product that succeeds in the market place and makes a profit for the company. The team should encourage critical thinking by the reviewers. They should try to understand all criticisms and should seek creative approaches for how they might be addressed.

7.7 DESIGN GUIDELINE

The purpose of the design guideline is to impose discipline and rigor on the design process. When properly formulated, it becomes the common design approach that is shared by all team members and that enables the team members to work together in an effective and productive way. The design guideline is the road map for performing product design within the company. It specifies a general design process to be followed, provides a variety of checklists that can be used for preparing for design reviews and for performing concurrent engineering, and recommends structured design methods and practices to be used at various stages in the design process.

Typically, the design guideline developed by a particular company is unique to that company. This is because each company has different business needs that must be met as well as different physical, organizational, and cultural constraints that must be satisfied. In addition, experience has shown that a design guideline is useless if it isn't actually used by the people performing the product design. As a result, it is usually found that the most effective design guideline is one that is developed by the people who are expected to use it. One of the best things a company can do to improve its design process is to give its employees the time and opportunity to develop a design guideline that works for them. It is also important to remember that, just like a product, the design guideline must be iteratively improved and refined over time and with experience. It must also be continuously adapted to changing needs and practices. The design guideline is therefore not a rigid "set of hoops" that the design team must jump through, but rather it is a living set of best practices that is continuously improved and optimized.

The formal design process is the heart of the design guideline. The review based design process shown in Fig. 7.6 illustrates one possible formal design process. This example illustrates several desirable characteristics. The process begins with the formation of the team. The core team is picked first and a team leader is identified. A core team member is also selected as a team recorder or secretary so that project records are kept and made part of the product design history and project file. The core team then works with management to select the extended team.

The project begins with a "kick-off" meeting, which is attended by all team members. The purpose of the kick-off meeting is to set the aspirations for the project and to communicate the overall "game plan." It is an opportunity for all team members to become familiar with the project scope and purpose. In a well planned kick-off meeting, marketing will present an overview of the market, the product opportunity including market size and target price, key product requirements and performance goals, and a review of competitor products. Engineering will typically discuss the state of technology including any development already completed or contemplated. Various design concepts and possible alternatives may also be discussed.

Formal design reviews can be time consuming and taxing on the team. For this reason, it is recommended that the team conduct numerous internal design reviews, but the number of formal reviews should be held to a minimum. The review-based process of Fig. 7.6 utilizes three reviews, which seems to be an appropriate number for many projects. The focus of the marketing review is to ensure a "winning" product from the external customer's point of view. Issues reviewed include customer requirements and product specifications, product concept, marketing and sales plans, and performance and design for manufacture issues. Key questions focus on the design alternatives considered and the reasons for selecting the proposed concept. Market potential, target sales price, production volumes, and other related issues are also discussed.

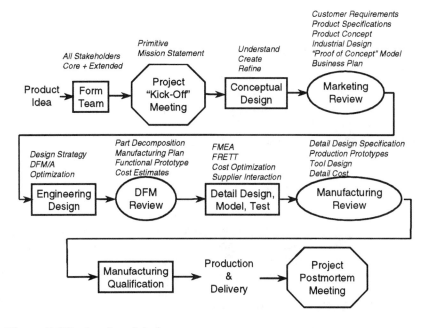

Figure 7.6 Review based design process.

The design for manufacture (DFM) review is critical for many product designs because the issues reviewed bear directly on ease of manufacture, life cycle support, cost, standardization, and ultimately, product profitability. The primary focus of the review is on the part decomposition and estimated cost. Issues include the assembly plan, appropriateness of part configuration, material selection, and manufacturing process selection.

The manufacturing review is held when the team members believe they have achieved the project's goals. The objective is an in-depth technical review of the design including detail bill of materials and cost analysis. Emphasis is on manufacturing and tooling issues, supplier issues, and other manufacturing and assembly concerns.

The responsibilities of the team are not completed until the product is tooled and in production. At some point following initial production ramp-up, the team conducts a "postmortem" meeting to review the project, discuss "lessons learned," discuss process improvement possibilities, and officially close out the project. Although not necessary for the product design, the project postmortem can be an extremely useful exercise for continuously improving the design process. By formally remembering what worked well and what didn't, the team sets the stage for the next project.

Depending on the company and type of products involved, the design guideline will usually include much more than the formal design process. For example, it might also include checklists for conducting each phase of the project, for preparing for design reviews, and for guiding individual team member contribution. It may also recommend the use of structured design methods such as those discussed in this book. Also, the design guideline can suggest best practices and provide guidance regarding the use of design tools such as CAD/CAM/CAE software packages and other design and analysis software. Ultimately, the design guideline can be whatever the firm and its employees want it to be.

7.8 KEY TAKEAWAYS

- Iteration is a learning process that is an inherent characteristic of the design process. Unfortunately, iteration cycles caused by poor design decisions can lead to engineering change. Engineering change is highly undesirable because it often forces additional changes due to the ripple effect and can quickly suboptimize the design. When performed late in the design, engineering change can also be extremely costly and time consuming. Engineering change is avoided by improving the quality of design decisions.
- Quality design decisions require availability and systematic consideration of all relevant design information. This is facilitated by good communications and design discipline. Therefore, to improve the design process, the firm must improve communications and design discipline.
- One prescription for design process improvement is to use a team approach coupled with formal design reviews and a design guideline. When implemented properly, the team approach improves communication, formal design reviews impose discipline on the process, and design guidelines provide a common approach.
- One size does not fit all. Each firm must decide which of these practices they wish to use and how they wish to use it. Often, the best results are obtained when those involved on a day-to-day basis with product design are given the opportunity to develop an approach that works for them. Once implemented, the approach should be continuously improved.

8

Customer Focused Concept Design

8.1 INTRODUCTION

Concept development is the crucial first phase of new product design. The goal of conceptual design is two-fold: (1) identify, develop, and select the best physical concept for the product and (2) develop a preliminary part decomposition. By *best physical concept*, we mean the physical concept for the product that best satisfies customer needs. By *preliminary part decomposition*, we mean that the part decomposition is tentative and is only defined to the extent necessary to select the best physical concept, construct proof-of-concept prototypes, and to investigate and demonstrate feasibility of various product and construction ideas. Component definition is incomplete from a manufacturing point of view and the decomposition is not necessarily based on any preconceived design strategy. Hence, one or more follow-on design iterations may be necessary to define the most suitable part decomposition from a producibility and internal quality point of view.

In this chapter, we focus our attention on identifying the best physical concept for the product. Recall that the *physical concept* is the embodiment of the working principle or solution concept employed to provide the desired functionality of the product. Product characteristics established by the physical concept include the main functions and workings of the product, the features which make it desirable to own and use, the styling that makes it aesthetically pleasing, and the attributes that make it reliable, durable, serviceable, and efficient to operate or use.

The physical concept essentially defines the interface between the customer and the product. When highly successful products are reverse engineered, almost invariably it is found that the physical concept employed is very effective in implementing this interface in ways that are either unique or clearly

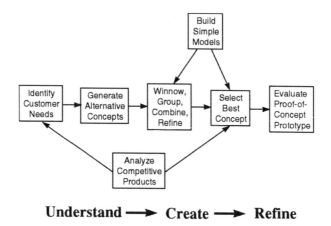

Understand ⟶ **Create** ⟶ **Refine**

Figure 8.1 The understand, create, and refine methodology involves researching customer needs to develop a customer focused design specification, generating alternative concepts, and evolving and selecting the best concept for satisfying customer needs.

superior to competitor products. This observation defines our guiding principle for customer focused physical concept design. That is, a *winning product* is one that (1) satisfies customer needs in the most optimal way and (2) is superior to all competitor products with respect to all important product characteristics. Our goal in this chapter is to present a simple methodology for defining winning new products.

8.2 METHODOLOGY

Developing the best physical concept for a new product is achieved using a three-step methodology (Fig. 8.1):

1. **Understand:** identify and prioritize customer needs and then translate them into a customer focused design specification.
2. **Create:** explore the design space by generating numerous ideas and concepts that meet or exceed the customer needs.
3. **Refine:** evolve the best alternative design concept by systematically evaluating and improving the ideas and concepts generated in step 2.

These three activities are at the heart of most successful new product developments. The *understand* step is pivotal because it tells the team what attributes and technical characteristics the product must have to succeed in the marketplace. The goal of the *create* step is to greatly increase the probability of identifying the best concept by generating a large number of alternatives from which to choose. In the *refine* step, the winning physical concept is evolved by systematically winnowing, grouping, combining, and refining the ideas and concepts until only a few highly promising alternatives remain. The best concept is then selected based on a systematic evaluation with respect to key customer needs and functional characteristics.

8.3 UNDERSTAND

Performing the research needed to identify, understand, and prioritize customer needs involves expressing the product idea in terms of customer needs and then associating metrics and target numerical values with each need. When formulated in this way, it is possible to use customer needs to guide both creativity and concept selection. *Quality function deployment* (QFD) is one of the best techniques for understanding and relating customer needs and design specifications. It can be applied to the entire product design or to any subproblem. QFD will typically require a significant time commitment. However, experience has shown that the benefits of using QFD are well worth the effort. The QFD process involves the following steps:

1. Formulate a statement of the basic design problem to be solved.
2. Determine the customers of the device or product.
3. Generate customer needs. What are the demands and wishes?
4. Develop engineering design specifications. How will customer needs be satisfied and measured?
5. Evaluate existing products or solutions. How is it done now?
6. Set targets. How much is good enough? What would be ideal?
7. Organize and evaluate research results.

8.3.1 Step 1: Develop a Problem Statement

Prior to beginning the research phase, it is important to clearly formulate a problem statement for the design effort involved. This establishes the level of the problem and forces the team to carefully consider exactly what the design task involves. The problem statement should include the main tasks to be performed together with carefully chosen focusing assumptions. *Focusing assumptions* are assumptions that limit the scope of the problem to a manageable level. For instance, by including the term "manually operated" in the problem statement, all means for powering the device other than manual are eliminated from consideration. Obviously, such assumptions must be chosen carefully since they have a strong influence on the types of solutions that are generated.

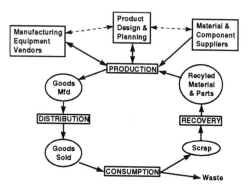

Figure 8.2 Each activity or component of the production-consumption cycle is a customer of the product.

In stating the problem, it is also important to constantly question the understanding of the problem to insure that the right problem is being solved. It is also extremely important to state the problem in a way that does not imply a solution or misguide the problem solving process. How the problem is defined determines the solutions that are developed. High quality solutions require high quality problem definition.

8.3.2 Step 2: Identify Customers

Most products have several types of customers. Who comes in contact with the product? Who purchases the product? Who uses the product? Who regulates the product? Who services and maintains the product? By systematically answering questions such as these, the team can identify the customers or customer types whose needs must be considered in developing a suitable design. For example, there are usually primary and secondary end users of the product as well as niche markets and ancillary markets. Each activity or component of the production-consumption cycle is a customer (Fig. 8.2). Similarly, each component of the product distribution channel (Fig. 8.3) and/or order-flow process (Fig. 8.4) is a customer.

8.3.3 Step 3: Determine Customer Needs

Customer needs for each type of customer are determined by interviewing individual customers, conducting and observing focus groups, sending out surveys, observing similar or competitive products in use, and so forth. Typically, the data that is gathered in this way must then be interpreted and translated into concise statements of need.

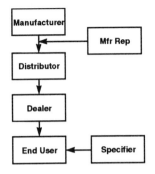

Figure 8.3 Each organization or individual in the product distribution channel is a customer of the product who has particular needs.

In a serious effort, it is not unusual to end up with a long list of need statements (50 to 300 is typical). Working with so many needs can be awkward. One way to deal with this problem is to group the need statements according to underlying similarities. These groups then form the primary list of customer needs that is used to guide the concept development process. To facilitate using the needs effectively, it is also often useful to develop a sense of relative importance of each need. This can be done based on consensus of the team or by using more elaborate methods such as further customer surveys or conjoint analysis. Since the correct decision regarding importance can, in some cases, make or break the project, this decision should not be treated lightly. Ultimately, the customer should verify all assumptions. Anything short of asking the customer could ultimately result in an unpleasant surprise.

If done correctly, identifying customer needs can be a time consuming and difficult task. Experience has shown however, that this step is extremely important and must be performed in a rigorous and systematic manner if a winning product concept is to be identified. Experience has also taught many useful lessons about what to do and what not to do. Ulrich and Eppinger (1995) suggest the following best practices for identifying customer needs.

- Get the whole team involved; don't delegate the needs process to the marketing representative.
- Understand the market, the technology, and the user.
- Observe customers using the firm's current product.
- Observe contextual behavior and best practices. Identify workarounds and other coping mechanisms. "Feel the pain" associated with using current products.

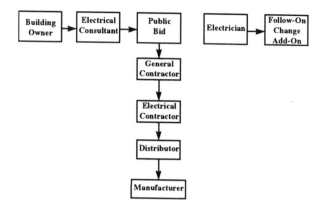

Figure 8.4 A typical order-flow for an electrical distribution product such as a lighting panel. Each organization or individual in the order-flow is a customer of the product who has particular needs.

- Capture the use environment by using video and/or still photography. These images teach a lot about user needs and can come in handy during concept generation and for making presentations.
- Identify *lead users* who experience needs ahead of others and who stand to benefit greatly from innovation or increased performance. These users can be a valuable source of customer needs as well as excellent judges for testing the validity of assumptions and ideas.
- Look for latent needs. *Latent needs* are needs that are currently unfulfilled or have not been clearly articulated or understood. Products that are the first to satisfy latent needs are often highly successful.
- Sometimes customers express preferences in terms of "how." When translating such preferences into needs, it is imperative that the essence of what the customer needs is captured, not how to accomplish it. When a useful solution is suggested, make note of it and save it for concept generation.
- Often, it is extremely useful to bring "props" to customer interviews and focus groups. Props help to stimulate discussion about needs and can often reveal problems. Useful props include an existing product, a prototype or mockup of an idea, or any one of many other items related to the use environment.
- Use matrices and other simple techniques to organize and prioritize customer need (Fig. 8.5).
- Never make assumptions about customer needs or their importance. The market will tell what it wants and doesn't want, just ask it.

Customer Type

	Type 1	Type 2	Type 3
Need 1	w_{11}	w_{21}	w_{31}
Need 2	w_{12}	w_{22}	w_{32}
Need 3	w_{13}	w_{23}	w_{33}
Need 4	w_{14}	w_{24}	w_{34}

Figure 8.5 The customer needs matrix relates the relative importance of customer needs to customer types. The weights w_{ij} assigned to each cell in the matrix reflect the relative importance of each need to each customer type. A variety of schemes can be used. For example, a "+" could be used to indcate that the need is important and "-" to indicate that it is unimportant. Or use a point scale of "1" for very unimportant to "5" for very important. The goal is to graphically display customer needs and customer types in a way that can be effectively used by the team to make decisions.

8.3.4 Step 4: Develop Engineering Design Specifications

The engineering design specification is the foundation of the product development program. It is usually stated in terms of technical requirements that can be measured and that express goals and limits for the product's technical characteristics. The design specification for the product will typically include requirements for performance, reliability, safety, economics, environmental effects, physical attributes (e.g., size and weight), aesthetics, manufacturing methods, life-cycle considerations (e.g., maintenance, service, and recycling), and so forth.

In QFD parlance, design specifications are referred to as engineering characteristics or attributes. *Engineering characteristics* are measures of the design solution's ability to satisfy customer needs. In developing the engineering characteristics, customer needs expressed in language of customers are translated into measurable attributes expressed in the language of the engineer. Consider, for example, the design of a heavy-duty mechanical pencil. Suppose that an important customer need is that the pencil be "easy to use." Engineering characteristics that reflect ease of use might include the length of line before refill, the amount of erasing before refill, the force required to produce a line of a given thickness or darkness, and adaptability to the hand. Note that some of these characteristics can be directly measured in a laboratory and expressed in engineering units such as millimeters, seconds, and Newton's of force. Others, such as adaptability to the hand, are subjective and are better measured using customer surveys or focus groups to rate the design using point scales or systems of "plus" and "minus" signs.

As illustrated by the mechanical pencil example, for most design problems, each customer need is effected by a variety of engineering characteristics. An important part of the QFD method is therefore one of winnowing and refining the engineering characteristics until a comprehensive and readily measured set is obtained. In general, only the most important and/or meaningful engineering characteristics should be considered.

8.3.5 Step 5: Analyze Competitive Products

What are the most important or meaningful engineering characteristics and what should their values be to insure business success? The first step in answering this question is to look at existing solutions to ascertain how the problem is currently being solved. The goal is to identify what is good and bad about current products and identify opportunities to improve. This is done by studying or *reverse engineering* existing solutions to understand their strengths and weaknesses from a marketing, technical, and manufacturing perspective.

In developing the Taurus and Sable automobiles, Ford Motor Company disassembled and studied more than 50 midsize cars from around the world in a methodical search for the best features of the best cars in the world. Ford found 400 "best-in-class" car features. At the time Taurus and Sable were introduced, the company said that the two models matched or bettered 80 percent of those features (Gerber, 1990). The practice of "reverse engineering" competitor products to find exemplary features which can be emulated or improved upon is a form of *competitive benchmarking*. The following steps are useful for benchmarking competitive products (Camp, 1989):

1. *Identify what is to be benchmarked.* In general, products can be benchmarked on three dimensions: (1) what customers like and don't like about the product, (2) performance with respect to key engineering characteristics (speed, weight, power, reliability, etc.), and (3) learning about detailed design features and approaches for economical production and ease of manufacture.

2. *Identify competitor products.* Finding the right standards to benchmark against is one of the great challenges of benchmarking. Often "best in class" competitive products are selected, but all design solutions have valuable lessons to teach. Expect to spend some time on this step.

3. *Determine data collection method and collect data.* Determine what information to collect and what measurements to use. In general, there is no one way to conduct a benchmarking investigation.

4. *Determine current performance "gap."* The team takes the measurements of its own existing and/or proposed new product concepts and compares them with the measurements of the benchmarked products. Differences determine negative, positive, or parity gaps that provide an objective basis on which to act.

5. *Project future performance levels.* The benchmarking team uses the measurement of the gaps to select the most meaningful or important engineering characteristics and set target values for the new product development program.

6. *Recalibrate benchmarks.* Once a physical concept has been selected and developed to the point where its performance can be evaluated, repeat steps 1 through 5 to verify the superiority of the new design. Because some engineering characteristics may be dependent on the particular physical concept selected, this step may also involve defining new engineering characteristics and redefining others.

Benchmarking by reverse engineering competitive products is largely a matter of common sense. To develop competitively superior products, the design team must know what the competition is doing. Most importantly, like Ford's development of the Taurus and Sable, the team must be open to learning from the competition and then be willing to apply this learning to the development of products that are better than the competition.

8.3.6 Step 6: Set Target Values

The results of Step 5 provide the insights needed to winnow and refine the engineering characteristics and set target values. Assigning target values helps guide the design and provide a basis for accepting or rejecting various solution ideas. It is often best to specify a marginal value and an ideal value for each engineering characteristic (Ulrich and Eppinger, 1995). The *marginal* value determines acceptability and is expressed as the minimum or maximum acceptable value for the engineering characteristic. When possible, marginal values should be precisely specified since they determine acceptability. The *ideal* value, on the other hand, is the value that would be most desirable or the best that can be hoped for. Ideal values can be specified precisely or in fuzzy terms such as "about" or "as low as possible" or "as high as possible." They can also be specified as ranges.

8.3.7 Step 7: Organize Research Results

Applying the quality function deployment method builds the QFD diagram or house of quality (Fig. 8.6). It is called a *house of quality* because the diagram resembles a house with many rooms. Each room contains valuable information that is developed by following the QFD process discussed above. In essence, the house of quality is a simple matrix showing the relationship between customer needs and engineering characteristics. As discussed previously, customer needs are usually qualitative while engineering characteristics are quantitative. For example, customers may want a lawn mower that is "easy to push and maneuver through the grass." One engineering characteristic might be the horizontal force to move through a three-inch high growth of rye grass.

It is desirable to have a clear relationship exist between the customer needs and the engineering characteristics. The relationship matrix (the bold box shown in Fig. 8.6) shows this relationship in the house of quality. The relationship matrix shows which and how much each engineering characteristic relates to each customer need. In general, each customer requirement should be measured by one or more engineering characteristics. In looking at Fig. 8.6, we see that engineering characteristic C2 does not measure any of the customer needs. This is an indication that this engineering characteristic is not needed and should not be used to drive design decisions. Similarly, customer need N3 is not measured by any of the engineering characteristics. This tells the team that an additional engineering characteristic is needed to capture this need.

The roof matrix shows interactions that may exist between the various engineering characteristics. For example, we see that improving C3 will also improve C1 and that the interaction between these two characteristics is judged to be strong. We also see that an improvement in C3 will negatively impact C4, but not too severely since the interaction is not strong. The roof matrix is useful in showing the tradeoffs and synergisms that exist between engineering characteristics. Relative importance of the various customer requirements and engineering characteristics helps guide the team in making tradeoffs and in deciding which features to emphasize and which features to sacrifice.

The house also provides a means for comparing particular design solutions with competitor products or existing solutions. The customer perception comparison shows how a new design solution compares qualitatively with existing solutions. In Fig. 8.6, each design is rated subjectively on a scale of 1 to 5, although any scale would suffice. In looking at how the new design compares, it is seen that it is somewhat deficient with respect to N2. Depending on the relative importance of N2, the team may choose to improve the new design to better satisfy this need. In theory, market success is ensured when the new design is perceived to be better than best in class competitor products with respect to all of the customer requirements.

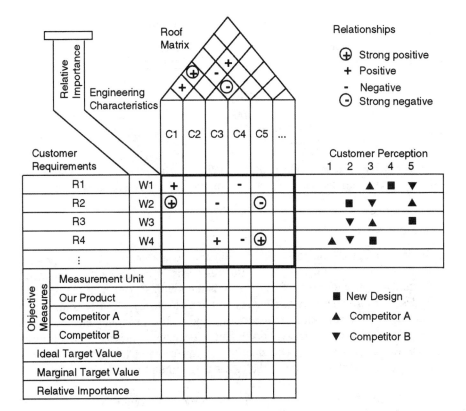

Figure 8.6 The house of quality.

The "objective measures" part of the house allows the team to benchmark new designs with respect to the engineering characteristics. This evaluation provides insight into how the designs compare technically. It also shows how numerical values of the engineering characteristics correlate with customer perception of the different designs. This can be very helpful in setting the marginal and ideal target values for the engineering characteristics.

There are no hard and fast rules for constructing and using the house of quality. The method is most effective when adapted to fit the particular product being designed. The goal is to clearly understand and define the product to be designed by relating qualitative customer needs to physically meaningful and measurable engineering characteristics. The house of quality does this by explicitly delineating the relationship between quality (satisfaction of customer needs) and function (engineering characteristics).

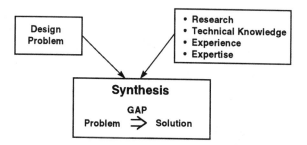

Figure 8.7 Synthesis is the joining together of the problem with potentially satisfactory solutions. The smaller the gap, the easier it is to build a bridge.

8.4 CREATE

In the create phase, numerous ideas and concepts are generated by the process of design synthesis to satisfy the customer needs disclosed by the understand phase. *Design synthesis* may be defined as the joining together of the problem of design with potentially satisfactory solutions (Fig. 8.7). The building up of solutions is accomplished by associating ideas. For example, the idea of a ballpoint pen leads to roll-on deodorant. Cockleburs sticking to wool socks leads to Velcro fasteners. The smaller the gap between the problem and the universe of potential solutions, the easier it is to build a bridge to the desired solution.

New ideas and concepts for products appear to be generated in three main ways (Fig. 8.8): (1) they are brought into existence by adapting or rearranging past or current design solutions, (2) they are brought into existence by imagination and ingenuity alone, or (3) they are brought into existence by using idea stimulating techniques. Adapting and rearranging existing solutions is one of the richest sources of new ideas. Existing ideas and concepts can be identified in a variety of ways. Some typical approaches include interviewing customers, benchmarking competitor products, examining patent files, visiting suppliers, and networking with "idea brokers" and universities.

The generation of creative design solutions by imagination and ingenuity requires the ability to get out of a "mental rut" and the ability to look at things in new and different ways. Important factors that contribute to creative ability include inherent or inherited mental abilities, knowledge, attitude, drive, and methods used. These factors are brought to the team by careful selection of team members and by disciplined use of "creative methods" or idea stimulating techniques. *Idea-stimulating techniques* are methods or practices that help stimulate the flow of ideas by narrowing the gap between the problem of design and potential solutions (Fig. 8.9).

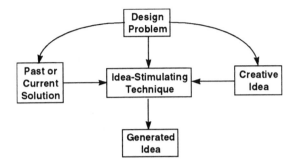

Figure 8.8 Sources of new ideas and concepts.

Idea stimulating techniques can generally fall into four different categories: (1) cognitive, (2) systematic, (3) analytical, and (4) constraint propagation. Cognitive methods involve mental exercises which help overcome rigidity in thinking and break through the implied boundaries of a problem. Synectics (Gordon, 1961) is a good example of a cognitive method. In this approach, taking different views of the problem generates new ideas. The different views include analogy, inversion, empathy, and fantasy.

Perhaps the best known and most widely used cognitive idea-stimulating technique is brainstorming. Developed by Alex F. Osborn (1979), the objective of brainstorming is to generate numerous ideas by having the participants focus on a specific problem during a "brainstorming" session. Brainstorming is most successful when the responses are uninhibited. The basic rules for brainstorming are few and simple (Raudsepp, 1983).

- Quantity, not quality, of ideas is the goal. The greater the number of ideas, the greater the probability that original and useful ideas will be surfaced.
- Defer judgment and evaluation of ideas until all ideas are in. This is the cardinal rule of brainstorming. Many people tend to evaluate and reject ideas while thinking them up. This is totally counter to the goals of brainstorming.
- Unusual or fanciful ideas are to be encouraged. Often, seemingly unfeasible, offbeat ideas turn out to be the best ones, or they may trigger ideas in others.
- When a group is involved, members should be encouraged to "piggyback" on the ideas of others, adding features or variations. This can increase the number and quality of ideas tremendously.

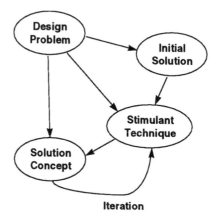

Figure 8.9 Idea-stimulating techniques help stimulate the flow of ideas by narrowing the gap between the problem of design and potential solutions.

Systematic methods seek to widen the search space for potential solutions by formalizing the process and making it explicit. The morphological matrix is a good example (Zwicky, 1969). This technique consists of three steps: (1) decompose the product or design problem into subproblems, (2) develop a variety of different solutions to each subproblem, and (3) visualize all the possible combinations of subsolutions by using a matrix.

Analyzing the problem mathematically can also stimulate creative ideas. In analytical optimization, for example, the problem of design is modeled mathematically in terms of an objective function and constraints expressed as functions of the design variables. An optimized design is achieved when the values for the design variables are such that the requirements of the design are satisfied and the objective function is minimized (or maximized). Often, however, one or more constraints limit how much the objective function can be minimized. When this is the case, the design team is stimulated to find creative ways to relax or eliminate the limiting constraints. In essence, the mathematical analysis shows what is possible and why it can't be realized with an existing or proposed design solution. By putting what needs to be done to improve the situation in clear and explicit focus, creative ideas and redesigns are stimulated.

In constraint propagation, downstream consequences of design decisions are propagated upstream to the early stages of design. Design for manufacture is a prime example. By considering manufacturability issues of a design early in the design process, undesirable quality and cost consequences can be avoided. Process driven design is another example. In this approach, the team decides how it wishes to manufacture and assemble the product before details of the design are decided. This imposes additional constraints on the design that force the generation of innovative design solutions. Constraint propagation is similar to mathematical analysis in that it alerts the design team to potential problems and in so doing stimulates creative design ideas for avoiding the problems, improving quality, and reducing cost.

There are many best practices associated with effective concept generation. Some of the most important of these are summarized as follows:

- Both complex and simple problems can be broken down into subproblems. Often, it is helpful to consider solutions to subproblems in addition to or instead of the whole problem. Exploring solutions to subproblems helps narrow the gap between the problem and solutions from other applications. It also facilitates using solutions from other applications. Note that problem decompositions are not unique. Often, several different decompositions may be found for the same problem.

- It is critical that the team look both inside and outside for design ideas. Brainstorming and other idea-stimulating techniques help the team look internally. To look outside, the team should consult lead users, patents, literature, competitive products, similar or related products, manufacturing personnel, and so forth to be sure the entire universe of possible approaches has been considered.

- Both individuals and groups generate good ideas. Therefore, to maximize the number of creative ideas, it is wise to seek a balance between individual and group approaches. Individuals generate ideas quicker than groups and each individual has a unique set of views and experience that he or she brings to the creative process. One of the best ways to tap into this potential as well as to prepare for a group brainstorming session is to devote time to generating ideas individually prior to the session. In addition to brainstorming, group sessions should focus on concept improvement and on building consensus that the solution space has been adequately explored.

- Set stretch goals, e.g., we want 400 ideas, 30 different concepts, etc.

- The best ideas are not necessarily the first to be thought of. Avoid becoming overly enamored with the first good idea that is generated.

- The understand, create, and refine phases overlap. Refinement during concept selection leads to new concepts. Customer reactions to new ideas lead to new concepts.

- Often, the physical concept for mature products is dictated by market or customer expectations. When this is the case, focus on finding creative solutions to subproblems.

8.5 REFINE

The goal of the "create" phase is to generate a large number of possible physical concepts for the product. The greater the number of candidate design solutions, the more likely the optimum or best candidate will be included. It is therefore exceedingly important that a large number of design solutions be generated. Using brainstorming and other idea stimulating techniques, a serious concept generation effort will typically result in hundreds of design ideas. The objective of the "refine" phase is to winnow and narrow the field of possible solutions in a very systematic and objective manner to ensure that the best possible solution is eventually evolved.

When selecting among a very large set of possible design ideas and concepts, the team should rely on common sense and on the insights gained in gathering customer needs and benchmarking competitive products. The following winnow, group, combine and select procedure is an effective general approach.

1. Winnow candidates that are clearly impractical or unsatisfactory for obvious reasons. Some may be impractical from a technical or cost standpoint while others may be unsatisfactory from a customer needs standpoint.

2. If many candidates remain, group candidates into natural sets having common characteristics.

3. Establish advantages and disadvantages of each candidate or group of candidates based on customer needs and development issues such as time and resources required. Winnow the less desirable groups and/or candidates within a group.

4. Combine the best features of the remaining candidates in each group and consolidate the group into a single (or a few) "best" candidate(s).

5. Formulate criteria for evaluation based on customer types, customer needs, and target values for engineering characteristics.

6. Use the evaluation criteria to optimize and refine the remaining candidates. Build simple models of the most promising concepts to verify functionality and to surface additional concerns and design considerations. Use the models to obtain customer reaction and then consolidate suggestions and improvements into new versions of the concept.

7. Select the optimal concept. One of many approaches is to evaluate each concept with respect to the "best in class" competitor product by assigning a "+" if it is better than, a "0" if it is the same as, or a "-" if it is inferior to the benchmark with respect to each evaluation

criterion. As a general rule, the selected concept will have the most "positives" and the least "negatives."

8. Reflect on the selected concept to be sure that it is, in fact, the right choice. Consider both the customer needs and the needs of the enterprise.

9. Manufacture a "proof-of-concept" prototype and test to verify that all target values for the engineering characteristics can be achieved and that performance and functionality are as expected.

Concept refinement and selection is an extremely crucial step in the concept development process. Selecting the right concept can be very difficult, especially in new product developments, where design information is very incomplete. And yet, this decision can eventually make or break the entire business venture so it must be taken very seriously. In the past, many firms have made very far-reaching concept decisions based on "gut feel" or management advocacy. This can be a very dangerous approach in today's highly competitive global markets and should be avoided. A much preferable approach is to utilize a disciplined process such as that outlined above. It is important to keep the following best practices in mind when performing the concept refinement process:

- Concept refinement is iterative and synergistic with the understand and create phases. Expect to discover unanticipated customer needs and design requirements during the refinement phase. Expect also to come up with new ideas and concepts as the result of concept evaluation. Keep an open mind. Constantly strive to combine the best features of different concepts into new and improved concepts.

- Concept refinement is hierarchical in that concepts are selected at the overall system level, the subsystem level, the component level, and the piece-part level. The methodology is essentially the same regardless of level. However, the effort applied should depend on the importance of the decision.

- Remember that any evaluation criteria not considered will not be included in the choice of the best concept.

- Always verify assumptions and uncertainties by asking the customer. Avoid "gut feel" decisions and decisions that are based on unsubstantiated customer needs. Avoid relying solely on opinions of engineers and others that are not every day users of the product. Recognize when a customer test is required to evaluate a concept.

- Beware of the best average product.

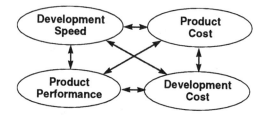

Figure 8.10 Each design project is usually "anchored" by one or more critical project goals or constraints. As a result, project anchors are often key selection criteria.

- Many design projects are "anchored" by certain over-arching project needs or goals (Fig. 8.10). These anchors interact in complex ways and can exert substantial influence on the concept refinement process. For example, low product cost may be incompatible with desired product performance and/or development cost and time. Satisfying both customer and project needs is often the most challenging aspect of concept refinement.

8.6 KEY TAKEAWAYS

Customer focused concept development is a process involving three distinct phases: understand, create, and refine. Understanding is required to establish customer needs, creativity is required to explore the entire design space, and disciplined refinement is required to evolve the best physical concept.

- Identifying customers of the product and correctly understanding their needs is the key to good concept development.
- Observing is a key product development activity. Observe customers using current products (both yours and the competition). Identify workarounds, best practices, "pain" associated with owning and using the product, and so forth.
- The quality function deployment (QFD) process is a highly effective structured methodology for understanding customer needs and defining customer focused design specifications. It can be applied to the entire product design or to any subproblem.
- Benchmarking competitor products provides insight into how to design a more desirable product that has better performance and is easier and less costly to manufacture.

- The first physical concept that is thought of is not necessarily the best. Therefore, a very large number of concept alternatives must be generated to ensure a high probability that the "winning" concept is included among all the alternatives generated.
- New ideas and concepts are created by adapting and rearranging existing designs, by using ingenuity and innovation, and by using idea-stimulating techniques. To generate a large number of distinctly different concept alternatives, each of these concept generation schemes should be employed in a disciplined manner.
- Innovation is a "team" sport and an "individual" sport. Employ a large range of idea stimulating techniques. Set stretch goals
- Concept refinement involves winnowing and selecting among alternatives. Discipline and objectivity is required to ensure that the best concept is eventually evolved.
- Most design projects are "anchored" by over arching project goals or constraints. These anchors often influence and/or dictate concept selection.
- The understand, create, and refine phases guide and inform each other through iterative and synergistic interaction.
- All assumptions and "judgement calls" should be customer tested. Customer needs and customer reactions are the guiding light for successful concept development.

9

The Rational Building Block Method

9.1 INTRODUCTION

The rational building block method combines the subdivide strategy with morphological analysis to form a systematic means for generating large numbers of design concepts. The basic approach is referred to by a variety of names such as the "concept combination table" (Ulrich and Eppinger, 1995), "systematic combination" (Pahl and Beitz, 1984), "morphological chart method" (Cross, 1994), and "technique for generating concepts from functions" (Ullman, 1997).

9.2 METHODOLOGY

The methodology is simple and straightforward. First the problem of design is decomposed into a set of simpler, more easily solved subproblems. Alternative solutions to each subproblem are then found and combined in different ways to create overall solution concepts. Solving simpler subproblems narrows the gap between the problem and potential solutions. Combining different subsolutions to create overall solution variants broadens the scope and helps to insure that all possibilities are considered. The method is implemented in five steps:

1. Decompose the problem of design into subproblems.
2. Identify and select the critical subproblems.
3. Develop alternative physical solutions for each critical subproblem.
4. If many subsolutions exist to a particular subproblem, narrow the list to the most promising and/or appropriate choices.
5. Develop design ideas by visualizing different combinations of the subsolutions using a matrix.

110

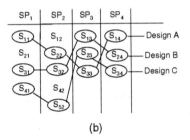

Figure 9.1 (a) Rational building block matrix. (b) Visualizing different combinations of the building block subsolutions generates design ideas.

The rational building block matrix consists of a columnar listing of possible physical solutions for each rationalized subproblem (Fig. 9.1a). Each element in the matrix represents a building block. Visualizing different building block combinations stimulates ideas for a variety of interesting and innovative design alternatives (Fig. 9.1b). Simple arithmetic shows that a 3x3 matrix (three subproblems, each having three solutions) results in 27 possible building block combinations. A 5x5 matrix results in 3125 possible combinations. Steps 2 and 4 of the methodology are necessary to keep the number of combinations from becoming overwhelming. More importantly, by rationalizing the subproblems and subsolutions, the idea-stimulating process is kept sharply focused on the heart of the design problem.

9.2.1 Functional Decomposition

There are many approaches for decomposing the problem. For example, we could subdivide the problem simply by analyzing it and deciding what the three or four major subproblems are. Or, if the problem of design involves a lot of user interaction, the design team could list the sequence of major user actions required and use these to identify the important subproblems. Still another

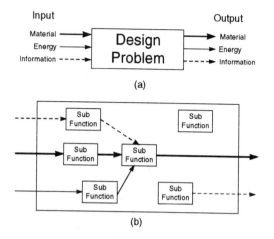

Figure 9.2 Functional decomposition: (a) "black box" representation, (b) resulting block diagram of functional structure.

approach might be to list the key customer needs involved and use these to identify the subproblems. This approach works well when the problem of design involves form rather than a working principle or technology.

In problems that are functionally complex, frequently the best approach is to decompose the problem in terms of the functions that must be performed. In this approach, which is called *functional decomposition* or *functional analysis,* the design problem is represented as a black box which converts given inputs into required outputs (Fig. 9.2a). The inputs are converted into outputs by a complex set of tasks inside the black box. These tasks are determined by breaking the design problem down into subfunctions. Each subfunction is expressed as a verb plus one or more nouns to help ensure that a particular solution is not implied. In general, each subfunction can usually be further divided into even simpler subfunctions. The end result is a block diagram that represents the design problem in terms of subfunctions, either connected together by material, energy, and information flows or shown separately without connection (Fig. 9.2b).

In original or very challenging problems of design, a good functional decomposition can be the key to finding truly creative solutions. Unfortunately, there is no really objective or systematic way for dividing a given design problem into subfunctions. Nor is there a single unique decomposition. An effective approach in many cases is to draft several different possibilities and then winnow, combine, and simplify until the problem is clearly and unambiguously decomposed.

9.2.2 Generating Subsolutions

The goal of step 3 is to generate as many solutions as possible for each subproblem. A large set of subsolutions greatly increases the probability that the entire region of feasible design solutions will be investigated and that the best or most appropriate overall solution will be synthesized. In addition to coming up with subsolutions by imagination and ingenuity alone, possible subsolutions can also be generated using the following techniques:

- Search for existing solutions
    - Investigate how similar problems are solved elsewhere.
    - Search for off-the-shelf components that may provide the desired subfunction. The *Thomas Register of American Manufactures* and the *Worldwide Web* are good places to look.
    - Search published literature. For example, Chironis and Sclater (1996) present an encyclopedic collection of mechanisms and mechanical devices for meeting a wide variety of product needs.
    - Search patents.
    - Consult experts.
    - Use software packages. As an example, the "IM-Effects" module of the *Invention Machine Lab* (Invention Machine Corporation, 200 Portland Street, Boston, MA 02114-1722) allows users to look up an effect by the function they want it to perform.
- Use creative techniques such as brainstorming.

9.2.3 Rationalization Process

The rationalization process performed in steps 2 and 4 should be conducted carefully to ensure that viable alternatives are not prematurely eliminated. Fortunately, for most design problems, identifying the crucial subproblems and most appropriate subsolutions is fairly straightforward. When there is a question, it can usually be resolved by mentally exploring possibilities or by team discussion. The goal of the method is to stimulate the generation of design ideas, not to be an overwhelming and tedious exercise. As is the case with most design methods, good judgment and practice are required.

9.3 QUICK-OPERATING FASTENER EXAMPLE

Use of the rational building block method is best illustrated by example. Suppose we are faced with the need to design a quick-operating fastener for an application which requires high preload of the assembled components as well as frequent disassembly and reassembly. The problem of design, created by this need is therefore to *design a manually operated fastening system that assembles and disassembles quickly and is also capable of developing high preload.* In this context, preload may be defined as an internal force developed by the fastening system components which acts to prevent separation of joined members and fastening system components under the action of external load.

9.3.1 Step 1: Decompose into Subproblems

The first step in creating a rational building block matrix is to functionally decompose the overall problem into a set of subproblems (Fig. 9.3). An effective functional decomposition strategy is to determine the operations required for the most dominant flow associated with the problem and then develop details for the other flows based on these operations. Doing this, we see that the material flow is the most dominant. To create the preloaded assembly, the separate components must first be assembled together and brought into initial contact. Sufficient preload must then be developed and locked into the joint to complete the assembly. These operations define the three material flow subproblems (Fig. 9.3b).

With the material flow subfunctions identified, it is apparent that fulfilling these subfunctions requires an energy flow in which energy is converted into force and movement. This leads to the "change energy" subproblem and energy flow shown. Similarly, the "indicate" subproblem and associated information flow shown is required to ensure that each material flow operation is correctly performed.

9.3.2 Step 2: Select Critical Subproblems

In developing the function diagram (Fig. 9.3b), we have deliberately tried to avoid an excessive number of subproblems by defining each subproblem at an appropriate level. Even with this care, five subproblems have been identified. This is more than we wish to consider given the degree of complexity of the problem. In looking at the function diagram, it is fairly obvious that the "indicate" subfunction is not critical, so we immediately decide to defer solution of this subproblem until later. A closer examination of the "change energy" subfunction shows that this subproblem must be solved differently for each of the material flow subproblems:

- "Assemble components" involves low force and large movement,
- "Generate preload" involves high force and small movement, and
- "Lock preload" involves low force and unspecified movement.

This realization leads to three critical subproblems, which we choose to focus on. These are shown in the context of the function diagram by the dashed lines in Fig. 9.3c.

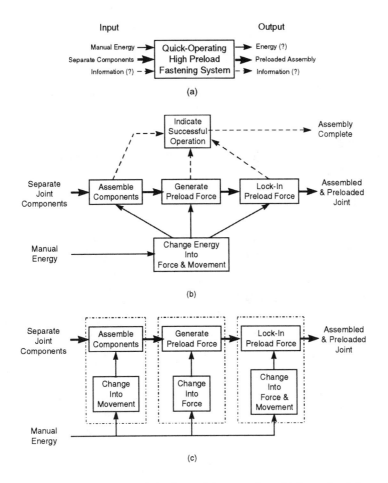

Figure 9.3 (a) Black box model of a quick-operating, high preload fastening system, (b) functional decomposition of fastening system, (c) rationalized function diagram with critical subproblems identified.

9.3.3 Step 3: Determine Possible Subsolutions

In this step, we propose subsolutions for each of the subproblems selected in Step 2. Considering the "assemble components" subproblem first, we propose the following categories of subsolutions: (1) multi-stroke, (2) single-stroke, (3) external, (4) snap-acting, and (5) automatic. These methods all involve changing manual energy into force and movement.

The multi-stroke category includes threaded devices, geared mechanisms, and other methods requiring multi-movements to produce motion. Single stroke subsolutions includes cam devices, slider-crank mechanisms, and other mechanisms that produce the required motion with a single stroke. External methods include all subsolutions that require manual manipulation, impact loading (hammer blows), simple tools or fixtures, and hydraulic or pneumatic systems. Snap acting includes mechanisms that involve elastic deformation and energy exchange as part of the assembly process. Automatic methods are those where assembly occurs automatically when the process is triggered. Pyrotechnic devices and other chemical or thermal processes as well as memory metal (Nitinol) mechanisms are included in this category.

Considering the "generate preload" subproblem next, we again see that there are a multitude of possible subsolutions. These include (1) the incline plane (e.g., wedge, tapered pin, screw thread, cam, etc.), (2) lever systems (e.g., toggle linkage), (3) external tool, (4) hydraulic or pneumatic systems, (5) gravity force (weight), (6) thermal expansion and contraction, and (7) electric or magnetic force (e.g., solenoid, electrostatics, permanent magnets, etc.).

Finally, four basic approaches for "locking in preload" are available. These are (1) the friction lock, (2) the mechanism lock, (3) the detent lock, and (4) the retaining lock. Friction locks utilize frictional resistance to maintain preload. Common threaded fasteners such as bolted joints are an example. Mechanism locks employ a mechanical linkage that can be put into a locked position. An "over-center" lock is developed, for example, by causing links to pass through the "toggle" position (position where adjacent links form a straight line). Once "over-center", the links are restrained from further rotation by a stop and are held against the stop by the preload force. In the detent lock, a catch, dog, or other spring operated holding device is engaged to lock in the preload. Retaining locks utilize separate components to lock in the preload force. Wired nuts or a stop, which prevents a wedge from backing out, are examples. The valve that locks in pressure and allows a hydraulic jack to support load is another example.

9.3.4 Step 4: Narrow the Choices

Many of the subsolutions determined in Step 3 can be eliminated based on limitations and constraints of the problem. Multi-stroke devices should probably be avoided because they would not be quick operating and the use of automated assembly does not make sense for a manual system. Similarly, only the incline plane, lever, and external tool fit the constraints of a simple, manually operated system. Finally, retaining locks should probably be avoided because of increased assembly and disassembly time as well as the additional complexity such solutions would require.

Assemble Components	Generate Preload Force	Lock In Preload Force
Multi-Stroke (AVOID) • Rotation • Translation • Other	Incline Plane • Wedge • Tapered Pin • Cam • Thread	Friction Lock
Single-Stroke • Rotation • Translation • Other	Lever • Slider-Crank • Linkage • Gears	Mechanism Lock
External Tool • Simple • Special • Impact	External Tool	Detent Lock
Snap-Acting		Retaining Lock (AVOID)

Figure 9.4 Rational building block matrix for systematic synthesis of quick-operating, high preload fastening systems.

9.3.5 Step 5: Generate Alternative Concepts

The matrix of rational building blocks can be constructed using the subsolutions developed in Steps 3 and 4 (Fig. 9.4). Note that some of the subsolutions eliminated in Step 4 are still included in the matrix but are designated as "avoids." By doing this, we minimize the importance of these subsolutions, but don't completely eliminate them from consideration. Each element in the matrix represents a building block. Ideas for a variety of interesting and innovative quick-operating, high preload fastening systems can be stimulated simply by visualizing possible building block combinations.

One way to verify the validity of the matrix is to see if existing solutions can be synthesized. To check this, we combine multi-stroke rotation, thread preload generation, and friction lock to generate the conventional nut and bolt fastener concept. Similarly, combining single-stroke rotation, cam preload generation, and detent lock leads to the commercially available quarter-turn fastener concept. Combining single-stroke translation, thread preload generation, and friction lock leads to a novel Sliding-Nut fastener, which is manufactured in Japan by the Mitsuchi Company. In this fastening system, movable spring loaded thread segments move outward allowing the nut to slide (translate) freely along the threaded portion of the bolt during assembly. Once snug, the nut is torqued in the conventional manner to preload the joint. These

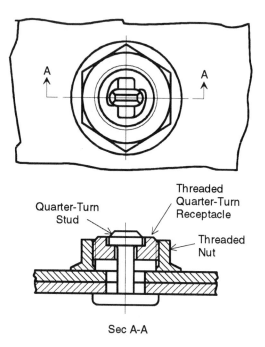

Sec A-A

Figure 9.5 Quarter-turn plus threaded take-up fastener concept.

examples convincingly illustrate that the matrix can be used to generate valid design solutions.

Examples of innovative new design solutions generated using the matrix are shown in Figures 9.5, 9.6, and 9.7. The quarter-turn plus thread take-up fastener concept (Fig. 9.5) combines the action of a quarter-turn fastener with that of an ordinary threaded nut (single-stroke rotation, thread preload generation, friction lock). The quarter-turn feature allows quick assembly and disassembly, while the threaded nut provides a means for accommodating tolerance stack-up and establishing joint preload by conventional torquing.

Figure 9.6 shows a variation of the quarter-turn plus take-up concept that utilizes a lever operated cam take-up in place of the threaded receptacle/nut. This concept is typical of the variations that are possible using different combinations of building blocks. In particular, a cam is used to develop preload in place of the thread and an "over-center" type mechanism lock is used instead of a friction lock. Note also that a keyhole is being used in place of the quarter-turn receptacle. The disc spring stack is used to accommodate tolerance stack-up and reduce the deflection required to produce the required preload.

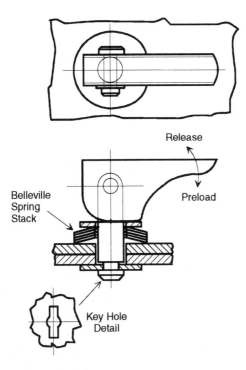

Figure 9.6 Quarter-turn plus cam take-up fastener concept.

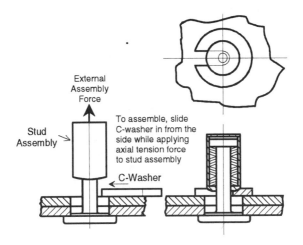

Figure 9.7 Sliding C-washer fastener concept.

As a final example, consider the Sliding C-Washer concept shown in Fig. 9.7 which utilizes an external tool to establish preload. The system consists of a bolt-like stud assembly that passes through the parts to be joined and a heavy C-washer that is used to complete the assembly. To install this fastening system, the stud assembly is first inserted through the parts to be joined. An external tool (such as a "pickle-fork" or hydraulic jack) is then used to apply axial tensile force to the stud assembly thereby compressing the disc spring stack. The C-washer is then inserted (translated) between the bottom of the disc spring stack and the surface of the parts to be joined. With the C-washer in place, the joint is preloaded because the disc spring stack cannot return to its original undeformed state when the external load is released. A detent type lock formed by the concave surface between the C-washer and the spring stack prevents the washer from accidentally slipping out. Disassembly of the system is the reverse of the assembly process.

9.4 DECOUPLED DESIGN

As discussed in Chapter 5, the root cause of many quality, manufacturing, and performance problems can be traced to undesirable interactions between various components or systems of a product design. *Undesirable interactions* occur when functional requirements become coupled in undesirable ways. One of the subtle, but extremely important benefits of using the rational building block method is that it tends to lead to "decoupled" design solutions. This is because the overall design solution is "built-up" from a combination of subsolutions that satisfy each functional requirement (subfunction) independently.

To illustrate, consider a conventional threaded fastener system and a conventional quarter-turn quick operating fastener system. In the case of the threaded fastener system, the multi-stroke motion used to assemble the components also develops the preload force once the components are snug. The infinite adjustability of the thread allows high preload force to be developed, but the multi-stroke motion involved in turning the nut makes assembly and disassembly unacceptably slow. In the conventional quarter-turn fastener system, the same quarter-turn movement used to assemble the components also generates the preload. Hence, it operates quickly, but isn't capable of generating much preload force. This is because the "assemble components" and "generate preload" subfunctions are coupled (are not functionally independent) in both of these fastening systems. These couplings (undesirable interactions) are acceptable as long as only one of the characteristics is desired, such as high preload for the threaded fastener or quick-operation for the quarter-turn fastener. They are unacceptable, however, when both quick operation and high preload are simultaneously desired.

In the case of the fastening system concepts generated using the rational building block method (Figures 9.5, 9.6, and 9.7), it is seen that the "assemble components" and "generate preload" subfunctions are satisfied independently. Hence all three designs avoid undesirable interaction between these functional requirements and are therefore capable of both quick operation and high preload. Based on this, we conclude that use of the rational building block method tends to result in "decoupled" design.

It is important to note, however, that the rational building block method does not guarantee a decoupled design. To illustrate, consider the quarter-tun plus take-up design (Fig. 9.5). Looking closely at this concept reveals an undesirable interaction between the "friction lock" and "generate preload force" subfunctions that can result in loosening over time. This undesirable interaction occurs because the friction force is developed by the preload force and is therefore dependent on it. As a result, variations in the preload force due to fluctuations of external load or vibration can cause the friction lock to slip over time producing an unacceptable loss of preload.

It is interesting to note that the undesirable interaction between the "friction lock" and "generate preload force" subfunctions has been a perennial problem with bolted joints. As a result, retaining locks such as cotter pins are often employed as "backups" in threaded fastener systems where loosening can't be tolerated.

9.5 KEY TAKEAWAYS

The rational building block method facilitates systematic synthesis of large numbers of concept alternatives by combining subsolutions in different ways. The method is both highly effective and easy to use.

- Most design problems can be divided into subproblems. There is no unique problem decomposition, however.
- The method is particularly valuable when the problem has been successfully decomposed into subproblems.
- Some approaches for decomposing the problem include:
  - Identify subproblems by analysis.
  - List the sequence of major user actions required.
  - List the key customer needs involved.
  - Decompose the problem in terms of functions that are performed.
- Using the method helps to avoid undesirable interactions.
- Like all design methods, the rational building block method is more applicable is some situations than in others.

10

Formal Concept Selection Methods

10.1 INTRODUCTION

The marketing, manufacturing, field support, and business consequences of concept selection are numerous and far-reaching. Many of these consequences are either irreversible or very expensive to reverse once the design solution is accepted and the product is in production. Concept selection is therefore one of the most important and momentous decisions made during the design process. Unfortunately, design information at the concept selection stage is usually very sketchy and incomplete. It is exactly for this reason that a concept selection method is needed to help ensure that the best or most appropriate concept is selected. By formalizing the decision process and making it explicit, the concept selection method provides a structured means for gathering and organizing subjective opinion.

Many approaches to concept selection are used. These range from using "pro and con" lists, following intuitive feel, or having the decision made by a concept champion, to using customer surveys or structured rating schemes or building and testing prototypes. Of these, structured rating schemes provide the best combination of discipline, flexibility, and amount of time and effort required. Experience has shown that concept selection is too important to trust to haphazard approaches such as intuitive feel and yet, information is too incomplete to spend a lot of time and money on hardware models and engineering analysis, especially if there are many alternatives to consider. Structured rating schemes strike a balance between these extremes.

Structured rating schemes work by dividing the concept decision into many subordinate, easier to make decisions. These subordinate decisions are then combined to arrive at an overall decision. The general procedure for using a structured rating scheme is as follows:

1. Develop evaluation criteria.
2. Assign an importance weighting to each evaluation criterion.
3. Rate each concept alternative with respect to each evaluation criterion.
4. Compute an overall score for each concept alternative that can be used for comparison and/or ranking purposes.

In the following sections we present two alternative implementations of this procedure: (1) the utility function method and (2) Pugh's method. The utility function method is relatively rigorous and, as a result, can be quite time-consuming depending on the amount of design information available and the level of detail desired. Pugh's method, on the other hand, is simple to use and can be applied quickly. Pugh's method has gained popularity because of its ease of use and because it helps the design team focus on the most important concept selection issues.

10.2 THE UTILITY FUNCTION METHOD

In the utility function method, the total effectiveness or *utility* of each alternative is obtained by summing numerical scores assigned to the individual constituent evaluation criterion, each weighted or proportioned to balance its importance in the final result. An analytical expression for utility, U, is given as,

$$U_i = \sum_{i=1}^{n} w_i x_{ij}, \quad (j = 1, 2, \ldots m) \tag{10.1}$$

where U_j is the utility of the jth concept alternative, w_i the weighting coefficient or measure of importance for the ith evaluation criterion, x_i the point score of the ith evaluation criterion, n the total number of criteria, and m the total number of concept alternatives considered. Eq. (10.1) is called a "utility function" because the evaluation criteria are assigned values or scores on a point scale. In more analytical applications, a similar equation might be termed a "criterion function" or "objective function." Using utility scores measured on a point scale allows both quantitative and qualitative evaluation criteria to be compared together.

10.2.1 Evaluation Criteria

The evaluation criteria are derived from customer requirements, the engineering design specification, insights developed during concept generation, team discussion, and so forth. Typically, there are a large number of possible criteria to consider. The key is to identify and properly characterize those criteria that are most relevant to the decision. Too many criteria will usually make the evaluation task overly tedious. At the same time, there is the danger that an important criterion may not be considered if only a few criteria are considered. Spending the time and effort to develop a clear, reasonably independent, and complete set of evaluation criteria is therefore essential.

10.2.2 Importance Weighting

Importance weighting is needed because not all evaluation criteria are equal. The essential task is to determine the relative importance of the criteria. Is reliability twice as important as weight? Is serviceability half as important as function? For convenience, the weighting factors, w_i, are usually assigned values according to the relation,

$$\sum_{i=1}^{n} w_i = 1.0, \quad 0 < w_i < 1 \tag{10.2}$$

where n is the total number of evaluation criteria.

Relative importance weights can be assigned in various ways. One of the easiest is to apportion 100 points between the different criteria according to their importance. This is usually best done by the team as a group. The weights are normalized by dividing the number of points assigned to each criterion by 100. Alternatively, the importance of each criterion can be rated on a standardized point scale of 1 to 5 (or 1 to 10 if more resolution is needed). This approach is illustrated for a hypothetical example in Fig. 10.1a. Relative importance can also be based on a comparison of each criterion:

1. Compare each criterion with each of the others. In each comparison, decide which is more important. Assign a "1" to the more important criterion and a "0" to the other. If both criterion are judged to be of equal importance, then assign "0.5" to each.
2. Sum values assigned to each criterion, x_i, to determine its relative importance, r_i, and normalize by dividing each r_i by the sum of all the r_i's.

A tabular implementation of this procedure is shown for a hypothetical example in Fig. 2.12b. If a criterion is assigned all zeros, the criterion may be dropped from consideration or an alternative method for assigning importance weightings may be used.

Importance	Point Scale
Extremely High	5
Very High	4
High	3
Moderate	2
Less Important	1

Criterion i	Importance Rating	Weight w_i
1	4	0.20
2	5	0.25
3	3	0.15
4	1	0.05
5	2	0.10
6	5	0.25
Σ	20	1.00

$$w_i = \frac{(Rating)_i}{\sum (Rating)_i}$$

(a)

Criterion, i	x_1	x_2	x_3	x_4	Row Sum R_i	Weight w_i
x_1		0.5	0.5	0.5	1.5	0.25
x_2	0.5		1	1	2.5	0.41
x_3	0.5	0		0.5	1.0	0.17
x_4	0.5	0	0.5		1.0	0.17
					6.0	1.00

1 = more important
0 = less important
0.5 = equal importance

$$R_i = \sum (row_i)$$

$$w_i = \frac{R_i}{\sum R_i}$$

(b)

Figure 10.1 Hypothetical examples illustrating (a) importance weighting using a standardized point scale and (b) weighting based on relative importance comparison of the criterion.

10.2.3 Scoring Methods

Because information about the candidate design concepts is usually sparse, assigning point values to the various evaluation criterion x_{ij} which accurately reflect each alternatives capability relative to each criterion can be challenging. Usually, the first step in doing this is to decide on an acceptable point scale. Typically, a scale of 1 to 5 is easiest to use, however scales such as 0-8, 1-10, and so forth can also be used if more resolution is needed.

Four different evaluation schemes are typically used depending on the nature of the alternatives, the amount of information available, and/or the evaluation criterion involved. These are (1) utility curves, (2) subjective rating, (3) paired-comparison, or (4) reference-based evaluation. *Utility curves* map numerical measures onto the point scale and are used when the evaluation criterion involves "quantifiable" numerical relations such as stress, weight,

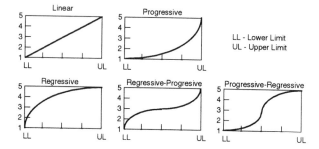

Figure 10.2 Typical utility curve shapes.

power, years of life, etc. Example utility curves are shown in Fig. 10.2. The *lower limit* represents the minimal acceptable performance level. The *upper limit* is set by the lesser of two conditions: (1) the best that can practically be achieved or (2) the most that can be effectively used. The shape of the utility curve determines the point values corresponding to particular numerical values of the evaluation criterion which fall between the upper and lower limits. The design team can pick from standard curves such as those in Fig. 10.2 or they can create new, distinctive shapes to reflect the exigencies of the particular design problem or evaluation criterion under consideration.

The other three scoring methods are useful when the evaluation criteria involve "judgment calls" or when there is insufficient design information to develop quantified measures. In the *subjective method*, scores are assigned by rating each alternative relative to each subjective criterion using a standardized evaluation scale such the one shown in Fig. 10.3. In the *paired comparison* approach, each alternative is compared with each of the other alternatives relative to a particular evaluation criterion and judged as to whether it is better, the same as, or worse than the other. Scores for each alternative are then calculated based on the results of the paired comparison. In essence, this approach reduces the evaluation task to a "which one?" choice as opposed to the judgmental "how much?" choice required by the subjective method. The procedure is similar to the relative importance weighting technique discussed previously. A hypothetical example is illustrated in Fig. 10.4.

The *reference-based method* uses a known reference such as an existing design or best in class competitive product as the standard of comparison. Each alternative is then assigned a point value for each evaluation criterion based on how it compares to the reference. Typically, a symmetrical 1 to 5 scale works best (Fig. 10.5). Note that a different reference design can be used for each evaluation criterion. This helps avoid scale compression when the reference is much better than or much worse than all or most of the alternatives.

Rating	Meaning
5	Complete Satisfaction: criterion satisfied in every respect.
4	Considerable Satisfaction: criterion satisfied in the majority of aspects.
3	Moderate Satisfaction: a middle point between complete and no satisfaction.
2	Minor Satisfaction: criterion satisfied in some but less than half the aspects.
1	Minimal Satisfaction: criterion satisfied to only a small extent.

Figure 10.3 Example of a standardized evaluation scale.

Alternative, j	A_1	A_2	A_3	A_4	Row Sum R_j	Point Score x_{ij}
A_1		1	0.5	0	1.5	2.50
A_2	0		1	0	1.0	1.67
A_3	0.5	0		0	0.5	0.83
A_4	1	1	1		3.0	5.00

1 = Better than
0.5 = Same as
0 = Worse than

$$R_j = \sum (row_j)$$

$$x_{ij} = \frac{P_{max} \ R_j}{n-1}$$

P_{max} = maximum point score

n = number of alternatives

Figure 10.4 Paired comparison evaluation. This hypothetical example illustrates evaluation of four alternatives ($n = 4$) with respect to a particular criterion using a five point scale ($P_{max} = 5$).

Relative Evaluation	Rating
Much worse than reference	1
Worse than reference	2
Same as reference	3
Better than reference	4
Much better than reference	5

Figure 10.5 A symmetrical five-point relative evaluation scale.

10.2.4 Ranking the Alternatives

Once the alternatives have been evaluated relative to each criterion, an overall score or utility value is calculated for each alternative using Eq. (10.1). Usually, the concept receiving the highest score is selected for further development. If the scores are close, the team might decide to retain more than one alternative

Evaluation Criteria	Weight	Alternatives			
		1	2	3	4
X_1	0.20	5 / 1.00	1 / 0.20	4 / 0.80	5 / 1.00
X_2	0.25	3 / 0.75	4 / 1.00	3 / 0.75	4 / 1.00
X_3	0.15	2 / 0.30	5 / 0.75	4 / 0.60	1 / 0.15
X_4	0.05	2 / 0.10	1 / 0.05	3 / 0.15	4 / 0.20
X_5	0.10	4 / 0.40	2 / 0.20	4 / 0.40	5 / 0.50
X_6	0.25	3 / 0.75	1 / 0.25	1 / 0.25	4 / 1.00
Σ	1.00	3.30	2.45	2.95	3.85

Figure 10.6 Decision matrix for hypotheical concept selection using the utility function method. Concept alternative 4 would be selected because it received the highest score.

for further development. Or, the team might decide to test the sensitivity of the decision by varying the importance weighting or individual rating values. In the hypothetical example shown in Fig. 10.6, the results are tabulated using a decision matrix. The decision matrix documents the process and makes it explicit. It also clearly shows how each alternative compares with respect to each criterion. By studying the matrix, the team can see which criteria, if any, are driving the decision. Sensitivity of the decision to importance weighting and evaluation criteria scores can be easily investigated by building a simple spread sheet model. The decision matrix also documents the concept selection process for future reference.

10.3 PUGH'S METHOD

Pugh's method (Pugh, 1991) is essentially a highly simplified version of the utility function method. The simplifications are summarized as follows:

1. All evaluation criteria are assumed to be of equal importance.
2. The alternatives are scored using the reference-based scoring method.
3. Instead of a point scale, concepts are scored relative to the reference using a "better than" (+), "same as" (S or 0), or "worse than" (-) system
4. The overall score is determined by simply counting the pluses, minuses, and "sames" for each alternative.

Evaluation Criteria	Concept Alternatives													
	1	2	3	4	5	6	7	8	9	10	11 (Ref)	12		
X_1	+	0	+	+	-	-	0	0	0	+	0	-		
X_2	-	-	+	+	-	-	+	+	0	+	0	-		
X_3	0	0	0	0	-	+	+	+	0	+	0	+		
X_4	-	-	0	-	-	0	-	-	0	-	0	-		
X_5	0	0	0	0	0	0	0	0	0	+	0	0		
X_6	+	0	0	0	0	-	+	+	0	-	0	-		
X_7	+	-	0	+	-	0	0	-	0	0	0	+		
X_8	-	-	+	-	-	0	0	0	-	0	0	+		
X_9	0	0	0	0	0	0	0	0	0	+	0	+		
X_{10}	0	-	0	0	-	-	+	+	-	0	0	0		
X_{11}	+	0	0	+	0	0	0	+	0	-	0	-		
X_{12}	+	0	+	0	0	-	0	+	0	-	0	-		
X_{13}	-	+	-	0	+	-	0	0	-	0	0	0		
X_{14}	-	0	0	0	0	0	-	-	0	+	0	0		
X_{15}	+	0	0	0	+	0	+	+	0	-	0	0		
Pluses	6	1	4	4	2	1	5	7	0	6	0	4		
Same As	4	9	10	9	6	8	8	5	12	4	15	5		
Minuses	5	5	1	2	7	6	2	3	3	5	0	6		

Figure 10.7 Hypothetical example illustrating Pugh's method. Based on the evaluation shown, Concepts 7 and 8 appear most desirable.

A hypothetical example of Pugh's method is shown in Fig. 10.7. Usually, one of the concept alternatives, preferably one that is well understood, is selected as the datum (concept 11 in Fig. 10.7). Creating the decision matrix (Fig. 10.7) focuses team discussion on the critical issues. This effectively generates the insight needed to choose the best one or few concepts quickly and without large amounts of detail design information. As in the utility function method, a decision matrix, such as that shown in Fig. 10.7, can be constructed as a spreadsheet model and saved to document the selection process. Pugh's method facilitates concept selection in a variety of ways.

- Scoring is simple. Deciding whether a candidate design concept is better than, the same as, or worse than a reference design with respect to a particular evaluation criterion is intrinsically easier than assigning a numerical score.
- The method is easy to use and to adapt as insight is gained. This makes it easy to repeat the evaluation process, possibly using different reference designs or evaluation criteria, as the selection issues and trade-offs involved become better understood.
- The matrix quickly shows which evaluation criteria are driving the decision. For example, in Fig. 10.7, neither X_5 nor X_9 appear to be very important. X_2, on the other hand, is producing a significant

influence. Similarly, the appropriateness of the reference design used for each criterion can be quickly ascertained. The scale compression occurring with X_5 is readily observed and can possibly be remedied by using a different reference design.

- The method makes it easy to evaluate a large number of alternatives using a variety of different evaluation criteria.

A disadvantage of the method is that it is not as rigorous or analytical as the utility function method. When both simplicity and rigor are desired, Pugh's method can be used as a rough screen to narrow the choice down to the best few. The utility function method, together with a more compact and orthogonal set of evaluation criteria, can then be used to select the best concept from the remaining few.

10.4 KEY TAKEAWAYS

- The effort applied to concept selection should depend on the importance of the decision.
- Structured rating schemes strike the best balance between intuitive approaches and expensive hardware evaluations. They are most appropriate when there are many alternatives to consider and design information is very preliminary and incomplete.
- The utility function method is appropriate for selecting among the two to five most promising alternatives. It is also useful when the evaluation involves quantifiable numerical relations.
- Pugh's method is effective and easy to use, especially when a large number of alternatives are to be considered. It is also useful as a screen for narrowing the choices to the few which are most promising.
- In using the utility function method, it is often useful to check sensitivity of the selection to importance weighting.
- Develop separate importance weightings for each different customer type. Perform concept selection for each different customer type and compare results.

11

Model-Driven Design

11.1 INTRODUCTION

Models are extremely important tools for improving the quality of design decisions and for taking time out of the product design cycle. In the early conceptual stages of design, simple physical models bring three-dimensional reality to design ideas. During configuration and parametric design of components and parts, computer-based solid models and computational design analysis tools allow a large number of alternatives to be quickly evaluated and help avoid costly and time-consuming hardware iterations. As the design is firmed up, the use of rapid prototyping methods facilitate timely construction of "looks like, works like" prototypes and allow tooling and manufacturing engineers to quickly visualize and react to part designs. In this chapter, we discuss various types of models and modeling practices that can be used to improve the product development process.

11.2 MODELS AND THE DESIGN PROCESS

Postmortem examinations of successful new product development projects often show that models play a key role at almost every stage in the development process. The new product development process typically involves a series of models that "inform" and "teach" by providing insight that cannot be gained in any other way, answering questions, and raising new questions (Fig. 11.1). The types of models used generally fall into two categories: (1) physical models and (2) computational models. Physical models bring three-dimensional reality to design ideas while computational models use the computer to simulate the design idea. The main purpose of both categories of models is to communicate design information and to provide design direction.

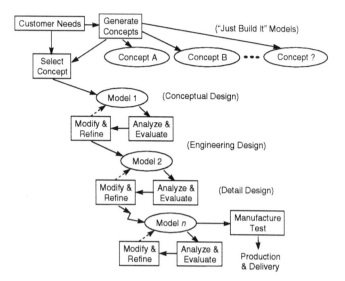

Figure 11.1 Model-based design process (adapted from McMahon and Browne, 1993).

11.2.1 Physical Models

Physical models are three-dimensional representations constructed from materials that may or may not be the same or have the same properties or exact form and structure as the final product or component. In general, there are four basic types of physical models: (1) appearance, (2) behavioral, (3) functional, and (4) design verification units.

Appearance models communicate how the product might look while *behavioral* models investigate how a design idea might be used or operated. Appearance and behavioral models focus on external features of the product and on the way the user interfaces with the product and interacts with it. They are useful for exploring different styling concepts and industrial design possibilities. They are also very useful in obtaining customer reaction and feedback.

Functional models are more concerned with the internal aspects or features of the product and with the way the product might actually work or perform its function. They frequently do not look anything like the final product and are often simple mockups or breadboards that demonstrate feasibility or that are used to investigate efficacy, performance, capability, and so forth.

A *design verification unit* (DVU) is used to validate or confirm the final design. DVUs typically look and work like the final product. In most cases, they are close or exact geometrical representations and are often made by processes that are the same as or that simulate actual production. DVUs are used to investigate tolerance stack-up, demonstrate performance, and obtain "looks like, works like" reaction and feedback from customers.

Physical prototypes can also be organized according to construction technique and time and effort expended. *Rough and ready* models are physical models that are constructed quickly out of readily available materials. They are generally built to answer questions or to develop insight into trade-offs and alternatives. *Design validation models*, on the other hand, are carefully made dimensionally accurate models. They are typically made by model builders using machining and other high fidelity processes. *Free form parts* are parts that are made using additive processes such as layered manufacturing methods in which the part is built up layer by layer. Several "rapid prototyping" techniques produce free form parts. Parts can also be made by removing material from prefabricated starting materials such as bars, plates, blocks, and so forth using either manual material removing methods or computer numerical controlled (CNC) processes.

11.2.2 Computational Models

Computational models are computer-based simulations that "approximate" the design idea. Computational models facilitate rapid investigation of many different alternatives. They can be used to investigate the kinematics and dynamics of complex machines and mechanical linkages, the tuning of an internal combustion engine, and the optimal design of a truss structure to cite just a few examples. One of the great strengths of computational modeling is the ability to model the relationships and interactions between geometry, material, process, and performance of complex physical systems. Stress, thermal, and flow analysis are now commonly performed using computational modeling. By computationally modeling the solidification of an alloy in a metal casting, the design of castings is rapidly being transformed from an art to a science.

Many computational models are based on "descretization" of component geometry. At the heart of most of these modeling approaches is the solid model. In theory, a *solid model* is an "informationally complete" geometrical representation of the design that permits well-defined geometric properties such as center of gravity, surface area, and assembly interference to be automatically calculated. Once defined, the solid model can be used to communicate form information to a variety of other applications including software analysis programs and rapid prototyping machines (Fig. 11.2).

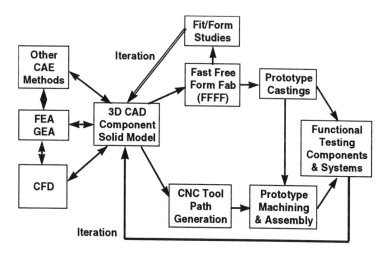

Figure 11.2 Computer-aided computational modeling and rapid prototyping environment (Conley and Stoll, 1995).

11.3 "JUST BUILD IT"

One of the quickest and most effective methods for understanding a design alternative and verifying its acceptability is to construct a physical model. Quick and dirty models answer questions, raise new questions, and provide insights that can't be obtained in any other way. Craig Sampson and John Lake of IDEO Chicago Product Development have recognized the importance of constructing physical models in the early conceptual stages of design and have termed this pivotal conceptual design activity the *"just build it"* philosophy of product development. In the words of Sampson (Conley and Stoll, 1995), "Fail early, fail often. Don't wait for all the information. What would you do if had the information? Enlightened trial and error always wins out over divine planning. Measure success by the height of the trash pile."

In the *"just build it"* approach, a proposed design decision is investigated by constructing a simple physical model or mockup of the design idea. To maximize effectiveness and timeliness, the model should be kept as simple as possible and should be made of readily available materials using readily available tools. Accuracy and precision are usually not big concerns. It is okay to build a sloppy model if it provides the answers and insights sought. The goal is to gain insight about a design decision before the details are defined to the point where appearance, accuracy and precision are important. By building a "rough and ready" physical model, the design team can experiment with ideas.

It is this experimentation which reveals the desirable and undesirable consequences and provides the insight needed to improve the decision.

To use the method, a small wood working and metal working shop, equipped with simple tools and benches and supplied with a variety of different construction materials (e.g., wood, foam core, foam, sheet metal, extruded aluminum shapes, PVC pipe, etc.) should be available. Without a readily available means for constructing simple physical models, the design team is likely to resort to less effective or reliable methods such as analysis or pure imagination. The shop should be a "hands-on" facility where members of the team can build and experiment. Complicating the process with paper work and employees hired just to build models defeats the purpose. The efficacy of the method lies in the fact that it gets the design decision makers directly involved with three-dimensional physical embodiments of their ideas early in the design process while there is still plenty of room to maneuver.

11.4 AVOID "TEST AND FIX" HARDWARE ITERATIONS

The design of many products, especially mature products, depends heavily on "art" rather than "science." Consider for example the design of circuit breakers and other electrical switching systems. Typically, these devices are designed using numerous "test and fix" hardware iterations because of the highly non-linear behavior of electrical arcs and the interaction of materials and geometry involved in the switching or current interruption process. Another example is the design of tooling for net shape manufacturing processes such as sheet metal forming and sand casting. The design and development of this type of tooling generally involves a "tryout" process in which an initial design of the tool is developed and fabricated. The tool is then tried and iteratively modified until acceptable production parts are obtained.

Problems with the "test and fix" tryout process are numerous:

- Many discrepancies are "mysterious." In these situations, there is typically no known solution and hence the only way to fix the discrepancy is by pure trial and error. This can often lead to suboptimal design solutions based on personal opinion and conjecture.
- The trial and error process is time consuming, unpredictable, and costly.
- Tryout only examines a narrow slice of possible variation. Once the product or tool is introduced into production, new problems and mysteries are often encountered because production involves reasonable variation of all variables. These must again be fixed by trial and error. No one knows why a change worked or even if it helped or worsened the overall situation.

- The process involves high risk. If all works out satisfactorily, a profit is made and the schedule is met. If serious problems are encountered, money is lost and the schedule slips.
- If tryout continues into production, expected part or product quality may never be achieved. In these cases, failure may not be admitted until it is way too late to be fixed. The result is unhappy customers and a cash drain that continues for the life of the product.
- In many firms, it is not unusual for "test and fix" design procedures to be surrounded with trappings of high technology. For example, a design solution may be proposed and then analyzed using computer-aided engineering (CAE) tools such as finite element analysis (FEA) or computational fluid dynamics (CFD). However, when carefully examined, it is seen that these are just digital simulations of the trial and error "tryout" technique – they can tell if a design solution is marginal but not how to fix it.

"Test and fix" hardware iterations are indicative of an immature core technology. Improving the design process in these cases requires the development of a "science base" for the core technology involved. Bridges are not designed by building a test structure and then driving heavy trucks over it to determine where to place the next strut. Rather, they are designed to perform their function by calculating loads and stresses using well-developed engineering principles of statics, dynamics, and strength of materials. Similarly, electrical wiring systems are not designed by trying different wires until one that does not burn up is found. Instead, each wire and component is sized by calculation.

A design *science base* is a well-developed and quantified, widely understood and appreciated, teachable, transparent, and broadly applicable design methodology. Typically, the science is "built-in" in the form of standards, codes, tabulated data, and well-developed training manuals. An effective methodology will provide a solid conceptual framework for problem solving, involve little or no use of higher mathematics, be fast, efficient, and unambiguous to use, and will result in acceptable design. Modern practices employed in the design of heating, ventilating, and air conditioning (HVAC) systems is a good example. Staring with a prescribed definition of customer requirements, the experienced practitioner uses a well-established design procedure, together with charts, graphs, and look-up tables, to quickly design an acceptable system.

Depending on the nature of the core technology involved and the level of awareness and capability existing within the firm, developing a design science base will involve some or all of the following steps.

1. Establish need by assessing time and cost of current process, problems, and opportunities for improvement.
2. Identify internal "champions" who will support and further the cause.
3. Identify sources and levels of knowledge inside and outside the company.
4. Develop needed knowledge. Work with academia and other R&D organizations if appropriate.
5. Organize knowledge into a problem solving design process.

   - Subdivide design process into procedures and steps.
   - Develop organized and systematized sets of activities, methods, and rules to be followed in each step.
   - Develop standardized knowledge building blocks such as look up tables, graphs, easy to use computer programs that converge quickly and unambiguously to the design result, and so forth.
   - Develop training procedures and train practitioners.

6. Ramp-up
7. Institutionalize

An effective science base and design methodology can be used to efficiently explore a wide range of different design options thereby eliminating costly and time-consuming hardware iterations. Most importantly, the science base and understanding provided by the design methodology makes it possible to optimize the design to a much greater extent than is possible using "test and fix" tryout approaches and practices.

11.5 CONSTRUCT PROTOTYPES AND DVUS QUICKLY

The advent of rapid prototyping (RP) technologies and bi-directionally associative three-dimensional CAD/CAM systems has had a significant impact on the speed and efficiency of the product development cycle. These evolving technologies facilitate remarkable development time compression and can lead to substantial product development cost savings.

Rapid prototypes serve many purposes. They can be used to verify form and fit and to check for interference between parts and ease of assembly. They also serve as marketing models and facilitate industrial design verification. They are also used in prototype parts procurement and as tooling masters. Most importantly, rapid prototypes provide rapid feedback to the design team and facilitate efficient decision making.

- **SLA** *Stereo Lithography*
  Company: *3D Systems Inc.*
  Materials: *epoxy*
- **SGC** *Solid Ground Curing*
  Company: *Cubital Solider*
  Materials: *wax, UV cured resins*
- **SLS** *Selective Laser Sintering*
  Company: *DTM Corporation*
  Materials: *PC, wax, nylon, and metal*
- **LOM** *Laminated Object Manufacturing*
  Company: *Helysis, Inc.*
  Materials: *paper, polyester*
- **FDM** *Fused Deposition Modeling*
  Company: *Stratasys*
  Materials: *wax, nylon, polyamide*
- **3-D Printing**
  Company: *Soligen, Inc.; BPM, Sanders*
  Materials: *Currently alumina/colloidal silica, ultimately any powder-adhesive combination*

Figure 11.3 Commercial rapid prototyping processes based on the additive approach.

There are three fundamental approaches for creating rapid prototypes: (1) additive, (2) subtractive, and (3) formative (Burns, 1993). In the *additive* approach, the part is fabricated by successively adding particles or layers of raw material to create a solid volume of the desired shape. In the *subtractive* approach, material is removed from a starting workpiece by CNC machining processes. In the *formative* approach, mechanical forces applied to the material form it into the desired shape. This includes both bending of sheet materials and molding of molten or curable liquids.

The first commercial additive process, called *StereoLithography*, was introduced by 3D Systems in 1987. Since that time, a variety of commercial processes have been introduced (Fig. 11.3). CNC machining has been widely used as a rapid prototyping technique for a number of years. At present, no widely used commercial process based on the formative approach has been introduced. However, there are hybrid machines that utilize the subtractive approach to shape sheet metal in the flat and then the formative approach to bend the sheet metal into three-dimensional shapes.

Obtaining rapid prototype parts that are functional can be a problem because of material and cost limitations. As an alternative to producing functional parts directly, masters produced using RP processes can be used to make tools that can produce functional parts. An example would be to use a rapid prototype master as a wax replacement to investment cast a functional part. Alternatively, the RP master can serve as the pattern for making a RTV silicone or cast epoxy mold which can then be used to make the wax pattern for investment casting. Still another alternative is to fabricate the wax injection tooling directly by using the RP process to generate a negative feature.

RP parts are generated using an STL file made from a solid model of the part. An STL file is created by slicing the solid model into layers, tessellating curved surfaces, and formatting the data in a standard form. Most commercial solid modeling software provides this capability. The steps for obtaining an RP part are as follows:

1. Create a solid model.
2. Generate an STL file.
3. Choose material and process.
4. Forward STL file to an appropriate service bureau.
5. Receive part in 1-5 days.

11.6 KEY TAKEAWAYS

- Models answer questions, raise new questions, and provide insight that can not be gained in any other way.
- Design information and insight is gained quickly by building simple models. The goal is to learn by "failing early and failing often."
- Innovation comes through experimentation.
- When using the "just build it" approach, don't worry about how the model looks or whether it works exactly right, focus on resolving uncertainty and establishing design direction.
- Product development teams seldom if ever regret building a model or prototype.
- Although building models is often the best way to understand and improve design decisions in new product development, it is not the best way to design new versions of existing products.
- Costly and time-consuming "test and fix" hardware iterations should be avoided whenever possible. Replace uncertain and unpredictable tryout design procedures with a science based design methodology.
- When design information is complete enough to create a solid model, obtain prototypes and design verification units quickly by utilizing rapid prototyping techniques.

12

Process-Driven Design

12.1 INTRODUCTION

Part decomposition involves the way the selected physical concept for a product is divided into subassemblies and parts. Traditionally, the part decomposition is developed by focusing on product function and piece-part cost. The resulting preliminary geometric layout and component designs are then iteratively redesigned until they are reasonably compatible with the method of manufacture to be employed. An assembly engineer then determines the best assembly sequence and designs the assembly process. This information is then used to develop a plan for mass producing the product including plans for fabricating tooling and producing or procuring the individual piece-parts.

Developing the part decomposition in this way can lead to cost and schedule overruns when manufacturing and other "downstream" requirements are not properly considered (Fig. 12.1). In this chapter, we present an alternative part decomposition process, which we call the process-driven design approach because it is essentially the reverse of the traditional approach. Instead of considering product performance and piece-part cost first, we start by developing a coordinated product and process plan that includes consideration of marketing strategy, manufacturing strategy, part standardization approaches, and method of assembly. We then develop a part decomposition that supports the coordinated product and process plan. The resulting design is further refined for ease of assembly and component fabrication. It is then iteratively optimized for performance, quality, and piece-part cost.

The process-driven design approach is as much a philosophy of design as it is a structured design method. The goal is to provide a framework for developing holistic product designs that maximize profitability by satisfying all customer needs, both external and internal.

140

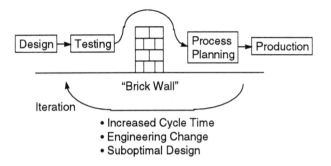

Figure 12.1 Developing a part decomposition without proper consideration of "downstream" requirements can result in increased design time, engineering change, and suboptimal design.

12.2 METHODOLOGY

The goal of the process-driven design approach is to ensure the development of the best possible part decomposition from a functional, manufacturing, support, and business point of view. This is done by systematically considering all "downstream" processes including life cycle support and enterprise requirements early in the part decomposition development. The process begins with an existing product or with the preliminary part decomposition and engineering design specification developed during generation and selection of the physical concept. The most appropriate part decomposition is then developed and refined using the following six-step procedure (Fig. 12.2):

1. Develop manufacturability design goals.
2. Develop a product and process plan.
3. Design components for ease of assembly.
4. Consider redesign of components for ease of fabrication.
5. Optimize and refine the design.
6. Iteratively improve as appropriate.

These steps are initially performed in a linear sequence. Once all aspects of the proposed part decomposition have been considered and impediments and problems identified, the part decomposition is further improved by iteratively reconsidering each step as appropriate. The objective is to quickly develop the best part decomposition from a conceptual point of view. Details are then further developed as part of detail design. A great advantage of the approach is that little or no major design change should be necessary during later stages of the design.

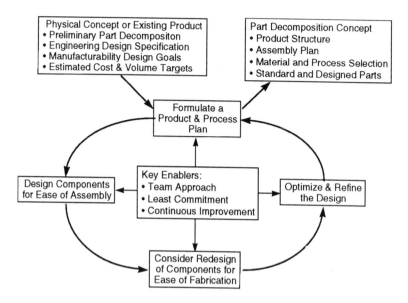

Figure 12.2 The process-driven design approach for optimizing and refining the part decomposition. Instead of begining with design optimization and then considering part fabrication and assembly (clockwise flow), the process-driven design approach considers the product and process plan together with ease of assembly and component fabrication, before optimizing and refining product performance and piece-part cost (counter clockwise flow).

12.2.1 Step 1: Develop Manufacturability Design Goals

A basic premise of the process-driven design approach is that product design is the first and most important manufacturing step. By planning the way the product will be manufactured and assembled before the detail product design is developed, the team is able to optimize the product and manufacturing process as a coordinated system. Manufacturability design goals provide a basis for manufacturing planning and for ultimately evolving the most appropriate part decomposition. The goals range from simple to complex. "Minimize the number of parts and part types" is a simple but highly challenging manufacturability goal. At the other extreme, a complex goal might entail detailed specifications for robotically installing particular subsystems of the product. For example, in the process-driven design of an automobile, a manufacturability design goal might be to "assemble and functionally test the rear axle and power train on separate assembly lines and then robotically install them into the vehicle at the last two stations of the final assembly line."

Sources of manufacturability design goals include the following:

- Customer needs.
- Engineering design specification.
- Manufacturing problems and manufacturing defects associated with production of the existing product or a similar product.
- Flexible manufacturing requirements.
- Seasonal fluctuations, high-mix low volume, and other business issues.
- Standardization and/or customization requirements.
- Product change and/or technology change.
- Reverse engineering of competitor products.

Manufacturability design goals for redesigning an existing product are developed by interviewing production workers, observing production and assembly processes, studying the way workstations and tooling have been iteratively improved to solve a production problem, and so forth. Marketing, manufacturing, and design strategies are also important sources of manufacturability design goals. For example, if a firm is planning to sell its product worldwide, designing the product so that the generic product can be customized to comply with varying electrical codes at regional distribution centers can be an important manufacturability design goal.

12.2.2 Step 2: Develop a Product and Process Plan

The product and process plan sets the stage and provides the design direction needed to quickly evolve the best possible part decomposition that implements the manufacturability design goals. In essence, the *product and process plan* is a plan for both decomposing the physical concept into subassemblies and components and for manufacturing and assembling them to form the final product. The goal is to identify the best part decomposition that (1) facilitates product performance, product change, product customization, product variety, component standardization, and product development management and (2) ensures that the components of the product are easy to manufacture and assemble with a minimum of quality risk and manufacturing complexity. The outcome is a proposed part decomposition plan that can then be analyzed and improved iteratively in steps 3-6.

Development of the part decomposition involves four major considerations.

- The division between standard and designed components.
- The product architecture to be employed.
- The assembly concept.
- Basic material and manufacturing process classes for key components.

Standard and Designed Components

Most products are combinations of standard components and designed components. A *standard* component is a physical element that is used interchangeably in a variety of different products and applications. Standard, "off-the-shelf" catalog components such as electric motors, light bulbs, electrical connectors, and mechanical fasteners, are referred to as *external* standard components. *Internal* standard components are standardized building block parts and chunks that are unique to a particular firm, but are used interchangeably in a variety of products and/or product families manufactured by the firm. A *designed* component is a unique part or subassembly that must be designed in its entirety during the product design. Decomposing the physical concept into standard and designed components is an important aspect of the product and process plan. For example, the choice between designing a special electric motor, optimized for performance and weight, or purchasing a standard motor from a supplier, can have far reaching performance, cost, quality, and timing consequences.

Product Architecture

Product architecture is the scheme by which the functional elements of the product are arranged into logical grouping of physical elements call "chunks" (Ulrich and Eppinger, 1995). *Functional elements* are the individual operations and transformations that contribute to the overall performance of the product. *Physical elements* are the standard and designed parts, components, and subassemblies that ultimately implement the product's function. A *chunk* is a collection of physical elements grouped together in a logical way to form a major physical building block of the product.

Chunks can be any logical grouping of components. For example, a chunk can be a simple part such as a metal stamping, plastic component, or purchased part. It can be a complex component such as a die-casting or insert injection molding. Chunks can also be formed by complex subassemblies such as a weldment fabricated from several parts, or a spot-welded frame, or a subassembly composed of designed and standard components. A chunk can be a purchased component or a stand-alone module. Chunks are usually formed to create a functional or manufacturing advantage. For example, chunks can be based on geometric integration and precision, function sharing, vendor capability, localization of change, accommodating variety, enabling internal standardization, portability, ease of service or maintenance, and so forth.

The following four-step approach for creating a product structure is recommended by Ulrich and Eppinger (1995):

1. Create a schematic diagram of the product showing the major functional elements of the product.
2. Cluster the elements of the schematic into logical groupings to form "chunks." Consider functional requirements, manufacturing and assembly requirements, serviceability requirements, customization requirements, and standardization requirements.
3. Create a rough geometric layout.
4. Identify and eliminate undesirable interactions.

A carefully planned product structure makes it possible to easily adapt, customize, and evolve the product to meet changing customer needs. It also provides the basis for standardizing various modules or chunks and for creating families of standardized building block parts and components. The ultimate benefits of a well thought out and executed product structure are reduced design time (e.g., only certain chunks need be redesigned) and reduced cost (e.g., standardized parts and chunks lead to economies of scope and scale).

Assembly Concept

The way the product components are assembled together to form the final product is determined by the *assembly concept*. The assembly concept includes the assembly structure and the assembly plan. The *assembly structure* relates to the way in which the physical elements are supported, oriented, located, joined, and integrated together to form the product as a whole. A variety of different assembly structures are possible. Some examples include a frame-based assembly structure, base component construction, stacked construction, building block construction, and so forth.

A *frame* is a structural unit that forms a skeleton on which the components of the product or device are mounted. Frames are typically composed of parts fitted and joined together. Representative examples include a bicycle or motorcycle frame, an automobile frame (e.g., space frame, unibody, rail, etc.), and an electronic rack assembly. In *base component construction*, one component (the base) is often considerably larger than the other parts in the assembly. In *stacked construction*, components are mounted in one or more layers on a base component and are often held together by a cover or top component (Fig. 12.3). In *building block construction*, which is typified by "Lego" block construction, different product variants are created by assembling different combinations of standardized parts (Fig.12.4). "Lego" block construction typifies the building block assembly structure. The assembly structures of many products are based on combinations of these and other construction approaches. Identifying the best assembly structure for the product is pivotal for achieving an easy to manufacture and assemble design.

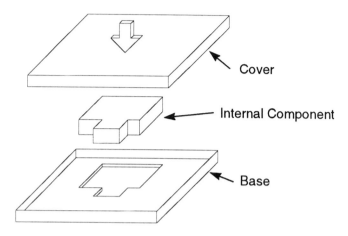

Figure 12.3 In a well-designed stacked construction, the product resembles a "z-axis" club sandwich. The internal components of the product are located and positioned in the base component and held in place by the cover.

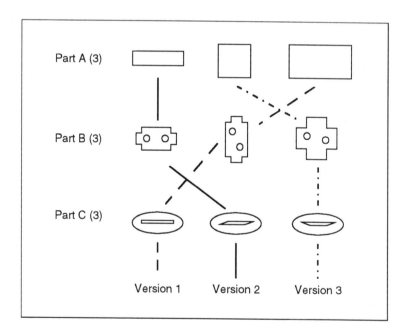

Figure 12.4 Theoretically, 27 different variants of this hypothetical product can be constructed using different combinations of the 9 building block parts shown.

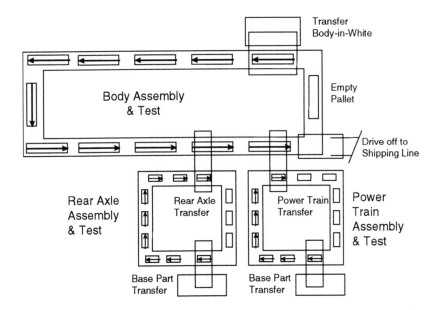

Figure 12.5 Plan for final assembly of an automobile that has been divided into three major chunks: body-in-white (frame), rear axle assembly, and power train assembly. Note that both the rear axle assembly and power train assembly are installed "bottom-up."

The *assembly plan* is the scheme employed for assembling the components together. It essentially specifies the sequence in which components (or chunks) are to be assembled and how the components (or chunks) are to be handled, inserted, retained, and inspected during assembly. To illustrate, consider an automobile that has been divided into three major chunks based on a "frame" based assembly structure: the body-in-white (frame), the rear axle assembly, and the power train assembly. One possible assembly plan for this assembly structure would be to load the body in white on the assembly line, assemble interior and exterior components, and then add the rear axle assembly and power train at the final two stations (Fig. 12.5). By including consideration of the product's assembly structure and assembly plan as an integral part of the product and process plan, assembly constraints and requirements are considered at the same time as performance and functionality.

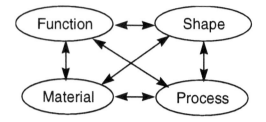

Figure 12.6 Material and process selection is governed by the complex interaction between function, material, process, and shape.

Material and Process Selection

Basic material and process class refers to the general material (e.g., metal, plastic, wood) and process (e.g., machining, sheet metal forming, near net-shape processing) to be used for the major components of the product. Depending on the type of product and design situation, these may or may not be determined or implied. For example, the exterior components of a typical passenger car are usually formed sheet metal. On the other hand, the enclosure of a new electronic device could be machined from a block of metal, formed from sheet metal, cast out of aluminum, or made of thermoplastic using a variety of different polymer processing methods. The actual material and process chosen depends on a number of factors such as projected total production volume, functional requirements (e.g., need for electrical shielding), and in-house manufacturing capability and expertise, to mention just a few.

Selecting the right material and process for major components can be a key consideration in developing a winning product and process plan. For example, part counts, assembly complexity, and secondary processing cost can be greatly reduced by integrating parts together using near net shape manufacturing processes such as plastic injection molding, investment casting, or powder metallurgy. However, the selection process can also be extremely complex because the material, process, and feasible part shape, size, and geometry are inextricably coupled (Fig. 12.6). Some considerations that make material and process selection difficult include:

- Many materials and processes to choose from.
- Must consider dozens of properties for each material.
- Must consider a variety of capabilities and limitations in relation to functional and production requirements.
- Must consider safety, cost, availability, codes, disposal, and so forth.
- Must consider part size, shape, and geometry.

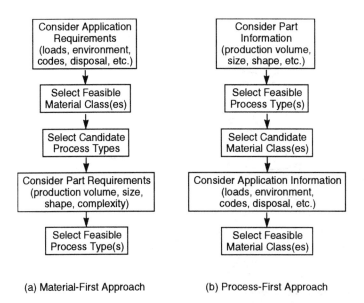

(a) Material-First Approach   (b) Process-First Approach

Figure 12.7 Alternative approaches for slecting material and process classes (Dixon and Poli, 1995).

Information is available however. The engineering design specification provides guidance regarding loads, environmental conditions, expected production volume, and so forth. Also, for most products, the general size, shape, and geometrical complexity of key components are understood on a qualitative basis. *Key* components are those components that have a direct bearing on product function or method of assembly. In addition, existing or available manufacturing capability and constraints are generally understood. One approach for selecting basic material and process classes is as follows:

1. Identify key product components.
2. Select feasible material classes and process types for each key component (Fig. 12.7).
3. Propose specific material and process alternatives.
4. Select based on the following criterion:
   • Product cost
   • Product performance
   • Development cost
   • Development time

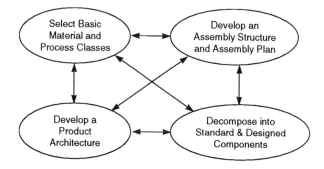

Figure 12.8 Identifying the best product and process plan involves systematically exploring the interactions and trade-offs that exist between the material and processes employed, the product architecture, the assembly concept, and the division between standard and designed components.

Explore Systematically

Identifying and developing the best product and process plan for the product typically involves trade-offs between the considerations discussed above (Fig. 12.8). How the process proceeds depends on the type of product and the design situation involved. If the product is totally new, the team may begin to plan a part decomposition by selecting the basic material and process classes. If, on the other hand, the product is essentially an assemblage of "off-the-shelf" parts, the team might start by selecting the standard components to be used, and then determine the assembly structure and unique parts that must be designed to integrate the standard components into the desired product. If the project involves the redesign of a mature product, product structure or the plan for final assembly might form the starting point.

A fundamental tenet of this book is that the likelihood of identifying the best design alternative increases dramatically with the number of alternatives considered. It is therefore important to consider a variety of alternative product and process plans before making a final selection. By generating and considering several different plans early in the design, the team is able to thoroughly explore the full range of design possibilities before constraints and interactions become tight. One way to systematically explore the feasible design region is to generate alternative product and process plans using a rational building block matrix (Fig. 12.9). The best product and process plan is then the one that best satisfies customer needs, avoids undesirable interactions, and minimizes information content of the manufacturing system.

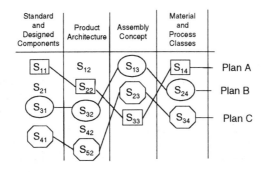

Figure 12.9 Generate alternative product and process plans by visualizing different combinations of subsolutions.

12.2.3 Step 3: Design Components for Ease of Assembly

In this step, part decomposition alternatives based on the product and process plan developed in step 2 are proposed, analyzed, and optimized for ease of assembly. The best alternative is then selected based on customer and project needs. The objective of this step is to eliminate parts (i.e., decompose the product into as few parts as possible) and design the parts that remain to be easy to handle and assemble.

In developing the redesign alternatives, the shape and geometric features of each chunk, subassembly and component as well as the way they assemble together are evaluated in terms of how well they conform to design for assembly (DFA) guidelines. Typically, the guidelines are stated as directives that act to both stimulate creativity and show the way to good design for assembly (see Chapter 13).

The following are design for manual assembly guidelines given by Boothroyd, Dewhurst, and Knight (1994). They are listed in decreasing order of importance based on the results of applying DFA analyses to a wide range of products.

1. Reduce part count and part types.
2. Strive to eliminate adjustments.
3. Design parts to be self-aligning and self-locating.
4. Ensure adequate access and unrestricted vision.
5. Ensure the ease of handling of parts from bulk.
6. Minimize the need for reorientation during assembly.
7. Design parts that cannot be installed incorrectly.
8. Maximize part symmetry if possible or make parts obviously asymmetrical.

Different and/or additional guidelines typically apply to robotic and automated assembly. In addition, most of the guidelines can be subdivided into an almost endless list of subguidelines that become more and more specific to particular applications and situations. For this reason, many firms have found it advantageous to develop distinct guidelines that apply more specifically to their particular products and fixed manufacturing facilities.

A major goal of this step is to maximize the probability of identifying and selecting the best part decomposition by generating a large number of different part decomposition alternatives. In Chapter 17 we present an easily applied qualitative manufacturability improvement methodology that can be used to systematically develop design for assembly redesign alternatives. Possibly the best known and most widely used design for assembly methodology is that developed by Boothroyd, Dewhurst, and Knight (1994). Although complex and time consuming to use, the Boothroyd-Dewhurst DFA Method provides quantitative results which enhances its usefulness. Various design for assembly methods as well as the development and evolution of design for assembly guidelines is also discussed in depth by Redford and Chal (1994).

12.2.4 Step 4: Design Components for Ease of Fabrication

The focus in this step shifts from the systems-level design of the product as a whole to the configuration design of each individual designed component. In component configuration design, the designer or design team is concerned with the impact of the general shape and geometric layout of the component on tooling cost, component quality, and lead time. For example, if the component is to be fabricated as a plastic injection molding, then it should be configured, if possible, to have a simple parting line, a minimum number of undercuts, generous draft, and liberal dimensional tolerances. The goal of this step is to ensure low piece-part cost by (1) designing the part to be easy to fabricate using simple tooling, (2) minimizing material cost, (3) minimizing processing cycle time, and (4) maximizing process yield.

Creating an acceptable component configuration typically requires that the preliminary size, shape, and geometric features of each component defined in Step 3 be modified as necessary to ensure compatibility with both the assembly plan and the specific manufacturing process to be used. To make this possible, the manufacturing process, and sometimes the material, must be specified in greater detail. For example, the choice of compression molding verses injection molding will depend on whether the material is a thermoplastic or a thermosetting plastic. Furthermore, the detail features of the part will depend on the nature and capability of the particular processing method and equipment to be used.

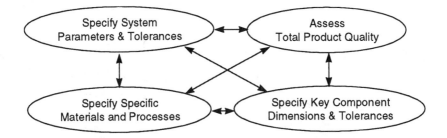

Figure 12.10 Major activities involved in optimizing and refining the design.

12.2.5 Step 5: Optimize and Refine the Design

This step involves four major activities (Fig. 12.10):

- Adjust and firm-up key system parameters to optimize performance.
- Select a specific material for each designed component.
- Specify detail properties and features of each designed component.
- Evaluate the design to ensure high quality.

Up to now, numerical values for key system-level design variables which determine product operating characteristics (e.g., speed, power, weight), product size and ease of use, and the dimensional relationships between components have been selected on a tentative basis. In optimizing the systems-level design, we seek numerical values for these design variables that maximize product performance, robustness, and manufacturability. Numerical values are generally firmed up and optimized through the use of engineering analysis and design optimization techniques (see Chapter 22), and, when necessary, fabrication and testing of breadboards and prototype models. The "just build it" design philosophy can be of great value at this stage to resolve uncertainties and verify optimality (see Chapter 11). Also, statistical problem solving techniques, such as design of experiments, can be used to determine the significant design variables and to establish values for these variables that maximize performance and/or minimize sensitivity to hard to control factors (see Chapter 22).

Optimization of the systems-level design provides the detail information needed to optimize the component design. The component design is optimized by (1) selecting the most suitable material and (2) specifying detail dimensions and tolerances (i.e., parameter design) which maximize performance and manufacturability and minimize total cost. Selecting the most suitable material involves identifying the material that offers the best acceptable combination of

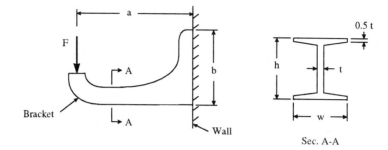

Figure 12.11 Bracket subjected to a cantilever-bending load.

low *total* cost, material properties (e.g., strength, corrosion resistance, temperature behavior, etc.), and compatibility with the particular manufacturing process to be used. Total cost includes the cost of the material plus the systems costs associated with purchasing, shipping, storing, handling, and recycling the material. Because so many factors and considerations are usually involved, it is strongly recommended that a disciplined material selection process be used. As a minimum, the process should include a formal requirement definition, generation of a list of candidate materials, selection recommendations, and a selection rational.

Parameter design involves specifying key component dimensions to maximize performance and manufacturability, and to minimize material cost. To illustrate this process, suppose a bracket, which is subjected to a cantilever-bending load, is to be fabricated as a plastic injection molded part (Fig. 12.11). Assume dimensions a and b and the force F are specified by functional requirements or as a result of the systems-level optimization. Assume also that the overall shape of the bracket as well as detail features such as the cross-section at Section A-A have been established based upon performance requirements, aesthetics, and design for plastic injection molding. In performing the parameter design, we wish to determine the detail dimensions of the bracket. To do this for the cross section at Section A-A, for example, we first determine dimension t based on the recommended nominal wall thickness for the selected plastic material. Using this numerical value for t and assuming there are no other constraints on h and w, we then determine numerical values for h and w which lead to the best combination of tooling cost, material cost, and allowable deflection and stress. If desired, we could further optimize the parameter design by varying h (or w or both h and w) along the length of the bracket so that the maximum bending stress is the same at every cross section.

Once the proposed design has been optimized on the systems-level and the component-level, the next activity is to assess the design from a quality standpoint. The purpose of the quality assessment is to ensure (1) that all undesirable interactions and potentially undesirable interactions have been identified, (2) all that could potentially go wrong with the product has been recognized, and (3) that appropriate actions are taken to avoid, prevent, mitigate, and/or accommodate undesirable interactions and potential failures. In addition, the assessment seeks to identify potential manufacturing and other life-cycle problems or consequences that may have been overlooked or may have resulted from the parameter design and design optimization.

12.2.6 Step 6: Iteratively Improve as Appropriate

Each step in the process-driven design approach is informed by and builds on the previous steps. As each step is performed, the design team learns more about the functionality of the design and about the way manufacturability, quality, cost, and functionality interact. Step 5 seeks to leverage this learning experience by providing a systematic means for incorporating learning experience into the design. This is done by repeating steps 1 through 5 sequentially as necessary until a globally optimal design is achieved. For example, it may be discovered in step 6 that the design could be further optimized or an undesirable interaction could be avoided if the particular part decomposition was modified in some way. In step 6, the team would make the appropriate modification and repeat steps 2 through 5 as necessary to harvest the identified improvement opportunity.

Using this cyclic iterative process, the design team is able to systematically optimize the functionality and performance of the part decomposition while simultaneously considering life cycle requirements. Because hardware is still remote and the design is still very fluid, iteration at this stage is relatively easy to perform. Also, by considering life cycle manufacturability requirements and constraints as part of each step and by performing the steps in a sequential and circular fashion, all available information is utilized and previous steps guide each step so the process converges quickly. By making iteration a planned part of the process, the process-driven design approach leverages the benefits of iteration while avoiding time consuming and costly iterations, such as those that can occur when major producibility problems are discovered late in the design after many irreversible hardware decisions have been made.

12.2.7 Key Enablers

Three key enablers facilitate the process driven design approach (Fig. 12.2): the team approach, least commitment, and continuous improvement. The team approach ensures that requisite product knowledge is available, when needed, so that all product requirements, including manufacturability and life cycle support constraints, can be properly considered. A policy of least commitment helps keep the design fluid so that the best compromises and tradeoffs can be made (Ostrofsky, 1977). An attitude of continuous improvement helps the team take advantage of opportunities discovered as they go through the process, even if it adds to project cost and time. The operative principle here is that a little extra time and effort spent early in the conceptual design stage is always more cost and time efficient than trying to harvest the opportunity at a later stage in the project or later in the product's life cycle.

12.3 OBSERVATIONS AND COMMENTS

The goal of the process-driven design approach is to provide the design team with a versatile, easy to use methodology for developing a part decomposition that systematically considers all manufacturing process and life cycle requirements of the product. The following observations and guidelines are presented to help provide insight into improving the effectiveness of the approach.

1. The process-driven design approach is based on the observation that a little structure and discipline goes a long way in facilitating good design. With this in mind, the team should not view this approach as a rigid step-by-step procedure, but as a philosophical framework for developing a part decomposition that systematically considers manufacturing process and other life cycle requirements. For example, when redesigning an existing product to improve performance or manufacturabiltiy, it is often more straightforward to combine steps 2 and 3 since the goal is to simplify assembly of the existing product. Remember also that, like all methodologies in this book, the process-driven design approach is a starting point for continuous improvement and refinement.

2. The approach is flexible and can be adapted to fit a variety of design problems. In general, each step is important and will always apply, but emphasis may shift depending on the particular needs of a given problem. For example, if the problem of design involves a totally new product, steps 3-5 may be of particular importance. If, on the other hand, a mature product is to be redesigned, developing the right product and process plan (step 2) may be the primary focus. The approach also works for subassemblies and individual components. However, in these situations, step 2 typically serves to clarify the design task while primary emphasis is placed on design for assembly and design for component fabrication.

3. For complex products (i.e., products that involve several modules or "chunks" or subassemblies), the process of formulating an effective product and process plan and design for assembly can often be speeded up significantly by rapidly cycling through the methodology in a highly focused and concentrated manner. One way for doing this is for the team to isolate itself in a meeting room (preferably off-site) with no interruptions until a basic part decomposition is developed. The goal is to rapidly identify several alternatives and examine the implications of each on a very preliminary basis. Because all team members are present and attention is sharply focused, the best approach can often be quickly identified from the choices available. If impediments are identified, team members can be assigned to develop solutions and these solutions can then be reviewed and alternative solutions brainstormed in a follow-up meeting.

12.4 CASE STUDY

Applying the process-driven design approach is best illustrated by an example. Figure 12.12 shows a schematic illustrating the functional elements of a typical window air conditioner. If one were to examine an exploded view of a traditional window air conditioner design, it would be seen that the final product consists of an assemblage of low cost, easy to make sheet metal parts. A typical bill of materials (BOM) might include the following:

* Purchased electrical and hardware components
* Two air-to-refrigerant heat exchangers (condenser for heat rejection to the outside and evaporator for cooling the room)
* A compressor for compressing and pumping refrigerant
* Copper tubing for the refrigerant circuit
* An expansion valve
* An electric fan motor
* A fan assembly for moving outside air through the condenser and a blower assembly for moving room air through the evaporator, each driven by the same fan motor
* Various enclosure and decorative parts

With the exception of the refrigerant compressor, fan motor, standard electrical components, and hardware, most of the parts are designed components. The large percentage of designed components is justified by the high annual production volumes typically associated with products of this kind.

The primary manufacturing problem associated with production of the traditional window air conditioner design involves refrigerant leaks. As

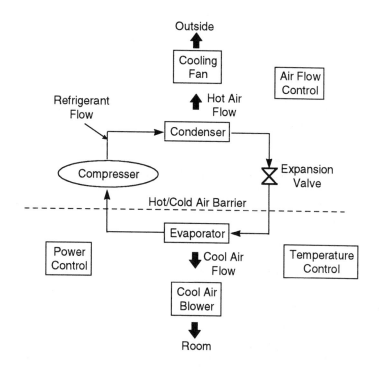

Figure 12.12 Schematic of a typical window air conditioner.

traditionally designed and manufactured, leaking units are detected during final testing of the completed assembly. Repair requires extensive disassembly, rework, and testing which is very time consuming and expensive. Other manufacturing problems include noisy operation (vibration, buzz, rattle, etc.), marginal cooling performance due to air leaks, and unsatisfactory appearance due to poor fit and finish. Many of these defects must be repaired before shipping. A further complication arises because of differences between the basic model and the upscale models, which have more appearance and control features. Typically, the line must be changed over or the different models must be assembled on different assembly lines. It is likely that microprocessor-based controls will make matters even worse in future models.

Let us imagine how this product might be redesigned for improved manufacturability and quality using the process-driven design approach. To do this, we will assume that the approach is implemented in a series of meetings involving the core team and others who are intimately familiar with marketing, purchasing, and manufacturing aspects of the product.

Meeting 1

Assume that the project is initiated by a meeting to discuss manufacturing problems being experienced by the current design and possible corrective actions that can be taken. Topics discussed in this initial meeting include a review of product background, discussion of current problems, and discussion of basic design for manufacture and assembly principles. This is followed by a brainstorming session focused on developing possible product improvement objectives. Key manufacturability design goals that are proposed include (1) refrigerant leak detection prior to final assembly, (2) automated final assembly requiring a minimum number of assembly workers, and (3) a significantly reduced part count. The meeting ends with agreement to meet off-site to develop a basic product and process plan that will achieve the design goals.

Meeting 2

This is the pivotal stage-setting meeting at which an overall product and process plan is developed. The meeting begins by sketching a schematic of the product (Fig. 12.12). The team notes that the compressor, condenser, expansion valve, and evaporator are connected together by copper tubing to form the refrigerant circuit. Since refrigerant leaks are a major concern, it becomes obvious after some discussion that the refrigerant system should be treated as a chunk. Also, since the control system is the one aspect of the product that is likely to change in future product generations, it is decided to isolate this potential design change from the rest of the design by making the control system a chunk. Finally, the fan motor, hot-air side fan, and cool-air side blower wheel are "chunked" together since both air-handling systems share the same motor. The possibility of using two separate motors is quickly eliminated when the purchasing representative points out that, next to the compressor, the fan motor is the most expensive component in the product.

   With the major chunks decided upon, discussion turns to the assembly concept. In considering an assembly structure, the team is strongly influenced by design for assembly guidelines and, as a result, decides to use a z-axis stacked construction approach. This leads to a basic product concept consisting of a base component and upper chassis component with the refrigerant system, control system, and fan system sandwiched in between (Fig. 12.13). In accordance with the design goals for leak detection and minimum possible number of assembly workers, the team proposes an automated assembly line concept (Fig. 12.14). In this concept, each chunk is assembled and tested on ancillary loops before being assembled on the main assembly loop. The team is enthused about this plan since, in addition to meeting the leak detection goal, it facilitates testing of all key systems prior to final assembly.

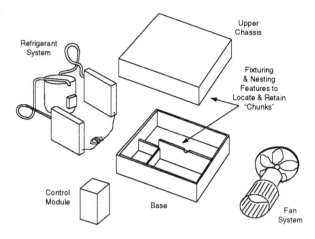

Figure 12.13 Product concept utilizing stacked construction and three major "chunks": the refrigerant system, fan system, and control module. Note that the outer wrapper (cabinet shell), which encloses the assembly, and, the front fascia (front assembly), which stylizes the unit, are not shown.

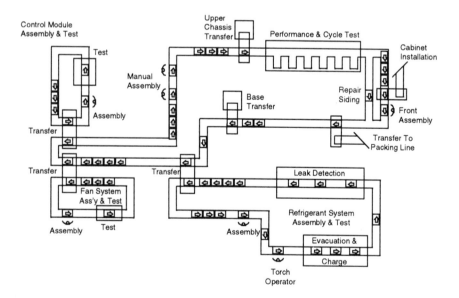

Figure 12.14 Conceptual process plan for the window air conditioner assembly. Note that each refrigerant system, fan system, and control module is fully tested before being robotically installed in the final assembly.

The next design aspect to be considered is possible material and process alternatives for the base and upper chassis components. These are considered to be "key" components because the assembly concept depends on being able to locate, fixture, and secure each chunk in the base and to form the airflow barrier and cool and hot air paths by providing mating features in the base and upper chassis. Because the team has considerable experience with sheet metal forming, it first attempts to come up with possible material and process alternatives by selecting sheet metal forming as the process type (process-first approach, Fig. 12.7). Consideration soon broadens to include consideration of casting and molding processes, however, since these are better suited for producing the required geometry with the fewest number of parts. Eventually, a proposal to make the base and upper chassis out of sheet molding compound (SMC) is selected as the most promising approach. There is considerable concern with this idea on the part of some team members because SMC is a radical departure from the company's core expertise of sheet metal forming. Also, all team members share a deep concern about material cost compared to the cost of sheet metal.

The next topic of discussion focuses on standard and designed components. Purchasing points out that it is strongly desirable to be able to purchase compressors from several different suppliers in order to be able to buy at the lowest possible price on the world market. However, to do this, differences in hole mounting pattern and envelope dimensions must be accommodated by the new design. The cost of the fan motor is then discussed and the desire to be able to purchase a smaller, less costly fan motor is identified as an additional key design goal. The possibility of purchasing "off-the-shelf" refrigerant-to-air heat exchangers is briefly discussed but quickly eliminated from further consideration since the firm has an extensive internal evaporator and condenser production capability. In essence, these items are seen to be "internal" standard parts since several standardized versions are widely used in a variety of products manufactured by the firm. Ultimately, it is decided that no major departure from the current division between standard and designed parts is warranted.

The meeting ends by assigning core team members with specific tasks. For example, product engineering is tasked with fleshing out and firming up the conceptual product design (Fig. 12.13), determining detail system geometry required for acceptable airflow and cooling capacity, and developing detail designs for the base and upper chassis components together with material and tooling cost. Similarly, manufacturing engineering is charged with firming up the assembly line concept (Fig. 12.14) and specifying end-of-arm tooling requirements for each robotic transfer station. Purchasing is charged with providing mounting details for all compressors it wishes the team to consider.

Table 12.1 Window air conditioner design improvement results

Old Design	Initial Product and Process Plan	Final Design
236 Parts	155 Parts	120 Parts
	Problem Areas • Air handling • Refrigerant system installation • Too many fasteners, seals, etc. • Complicated tooling • Flexibility and strength of base **26 Performance Concerns Identified** **43 DFA Weaknesses Identified** **9 Tooling and Process Concerns Identified**	**Major Improvements** • Elimination of producibility concerns • Improved performance – smaller fan motor • No fasteners in final assembly • Simplified tooling (all straight pull molds) • Hardware commonality • Elimination of seals and gaskets • Added guiding and nesting features • Unique compressor mounting plate **Process Modifications Required** • Sequence of assembly steps • Assembly line configuration • Coil and tubing connections relocated away from SMC base

Meeting 3

Meeting 3 is scheduled after the tasks assigned in meeting 2 are complete. This meeting has a threefold purpose: (1) review the product and process plan from a performance, design for assembly, and component design for manufacture perspective, (2) identify impediments, and (3) make assignments to team members to work out solutions to the various impediments. As a result of this meeting, a variety of impediments and concerns are identified (Table 12.1).

To illustrate the nature of these concerns, consider the refrigerant system installation. This system is composed of the compressor (approximately 18 lbs.), the evaporator and condenser (approximately 2 lbs. each), and expansion valve, all connected together by imprecise and springy coiled copper tubing. Because of the complex and unpredictable nature of this assembly, it was determined that the end-of-arm tooling required to pick up and install the refrigerant system in the base would be unreliable and extremely costly. In reviewing why this transfer was deemed necessary, the team realized that it had decided to build the refrigerant system on a separate line rather than using the SMC base as the fixture because of concerns regarding heat damage to the SMC base due to copper tube brazing.

Meeting 4

This meeting is held to review impediment solutions, brainstorm alternative solutions, and alter the product and process plan to remove the impediments. This meeting is repeated one or more times until all impediments have been removed and the design goals have been achieved to perfect the product and process plan.

Next Steps

Once the product and process plan are complete, the team then iteratively improves the design by performing steps 3-6 of the process-driven design approach. Often, it is convenient to hold meetings similar to meetings 3 and 4 to facilitate these steps. The results of using the process-driven design approach to improve the window air conditioner design are summarized in Table 12.1.

Project Postmortem

In reviewing how the process-driven design approach worked in facilitating an effective and efficient window air conditioner redesign, we note the following benefits and observations:

- All manufacturability design goals were met or exceeded. Experience has shown that setting "stretch" manufacturability design goals stimulates innovation and design creativity. Most importantly, by setting manufacturability design goals early in the conceptual stages of design they are almost always achieved, often in unique and unanticipated ways.
- The existing design had 236 unique part numbers. This was reduced by 81 parts to 155 in the initial product and process plan. A significant portion of this initial reduction was the result of the decision to mold the base and upper chassis instead of employing a built up sheet metal construction. Subsequent refinement and optimization of the design in steps 3-6 of the process-driven design approach resulted in the elimination of an additional 35 parts and 6 tools.
- By discovering the impediment associated with end-of-arm tooling for robotic transfer and insertion of the refrigerant system early in the

design, the team was able to avoid a potential problem during product launch. The product and process plan change was simple – combine the refrigerant system assembly line with the final assembly line and use the base component as a fixture for assembling the refrigerant system. This not only eliminated the need for a special fixture, it also eliminated the robot and end-of-arm tooling, and with it the cost and complications such a process would have created for the life of the product. To facilitate this manufacturing change, the tubing connections had to be relocated away from the SMC base and some detail geometry of the base had to be modified. Had this problem been discovered late in the program after the base had been designed, the solid model created, the finite element analysis completed, prototypes built and drop tested, and tooling ordered, it is very unlikely that product engineering would have been very open to the change. Instead, by using the process-driven design approach, this potential disaster was avoided by making minor changes to sketches of the base and upper chassis components.

- As part of the design for assembly analysis, several seals and gaskets were eliminated. These design changes allowed the two manual assembly stations on the final assembly line that were originally anticipated (Fig. 12.14) to be eliminated.

- In the "consider redesigning components for ease of fabrication" step, both the base component and upper chassis were redesigned to eliminate all camming and side action in the mold. This significantly reduced tooling cost as well as process cycle time.

- As part of the "refine and optimize" step, it was discovered that the design changes made to eliminate camming in the base and upper chassis molds also greatly improved airflow through the device. This provided the team with insights that eventually led to a geometry that allowed the team to achieve its goal of using a smaller fan motor.

- Material cost for the SMC base component and upper chassis was found to be greater than the cost of equivalent spot-welded sheet metal fabrications. However, because the information content of the SMC parts was considerably lower, the team was convinced that the direct and indirect cost saved by avoiding the fabrication and assembly of the built-up sheet metal components more than offset the additional material cost. Also, tooling lead times and cost for the SMC molds were significantly less than those for the several sheet metal stamping dies that were required. As a result, although there was some discomfort with the SMC material and process choice because it is not a core competency within the company, the team made the decision to recommend this material and process selection to upper management.

- A special adapter plate was designed to facilitate the interchangeable use of a variety of different compressor configurations without requiring any tooling or process changes. This would not have been possible had this need been identified late in the program.

12.5 KEY TAKEAWAYS

The process-driven design approach involves the systematic consideration of manufacturing process and other life cycle requirements and design goals during the early stages of conceptual design. The objective is to define the most manufactureable part decomposition as quickly and efficiently as possible while avoiding lengthy and costly design iterations.

- Conceptual design involves (1) identifying the best physical concept for satisfying customer needs, (2) identifying the most appropriate part decomposition for ease of manufacture, assembly, and life cycle support, and (3) integrating these into a design concept that optimizes total product value.
- The most appropriate part decomposition can be identified and optimized by considering assembly and component fabrication requirements prior to optimizing the design from a functional and piece-part cost perspective.
- The process-driven design approach provides a philosophical framework for developing the part decomposition and should not be viewed as a rigid step-by-step procedure. It provides the discipline needed to help ensure that all life cycle needs and constraints are properly identified and considered early in the design process.
- The process-driven design approach is applicable to products, chunks, subassemblies, and individual piece parts.

13

Part Elimination Strategies

13.1 INTRODUCTION

Elimination of separate parts is perhaps the single most effective way to eliminate information content. Parts account for the great majority of a product's cost, both direct and indirect. They are also the primary source of quality risk, unreliability, and customer dissatisfaction. Fewer parts mean less of everything involved in the design, manufacture, and support of a product:

- Engineering time and effort, design specifications, and part numbers.
- Form, fit, and finish details, tolerance stack-ups, etc.
- Quality risks inherent in the assembly process.
- Raw materials that must be specified, purchased, stored, and moved.
- Production control records and inventory.
- Purchase orders, vendors, receiving inspections, etc.
- Containers, stock locations, buffers, etc.
- Material handling equipment, part moves, part manipulations, etc.
- Accounting details, calculations, and records.
- Service parts, catalogs, training, etc.
- Production and assembly equipment, facilities, floor space, training.
- Packaging materials and component handling risks.

13.2 CANDIDATES FOR ELIMINATION

A part is a *candidate for elimination* (CFE) if there is no need for relative motion between it and parts already assembled, no need for subsequent adjustment between parts, no need for serviceability or reparability, and no fundamental reason requiring that materials be different. "Theoretical parts" are parts that can not be eliminated. For a part to be a theoretical part, it must receive a "yes" answer to *at least* one of the following critical questions (Boothroyd, Dewhurst, and Knight, 1994):

1. Does the part move relative to other parts?
2. Must the part, for good reasons, be made of a different material?
3. Does the part need to be separate for assembly or service?

To ensure that all redesign possibilities are explored, an answer of "yes" should only be assigned for fundamental reasons. For example, in order for a reciprocating piston internal combustion engine to operate, the piston must be able to move relative to the engine block. Hence, an answer of "yes" to the *motion* question is appropriate for the piston. An answer of "yes" for a spring, on the other hand, is probably inappropriate, even though a spring must move relative to other parts to function. This is because relative motion in the case of a spring usually involves elastic deformation, hence some part of the spring can often be integrated into a higher level part.

It might be argued that the spring must be made of a different material, perhaps because of fatigue or yield strength. For the purposes of the critical questions, however, mechanical failure is an engineering reason and not a fundamental reason. Similarly, the steel cylinder liner in an aluminum engine block is needed for engineering reasons not fundamental reasons. If an aluminum alloy with acceptable wear characteristics could be found, the separate steel cylinder liner could be eliminated. Hence, the steel liner is a candidate for elimination and an answer of "yes" to the *material* question for this part would be inappropriate. An answer of "yes" to an electrical conductor, on the other hand, is appropriate because clearly the need to electrically insulate a conductor from its surroundings is a fundamental reason for materials to be different.

The third question is often the most difficult to answer. Many parts need to be separate for reasons of assembly and service. Purchased components such as ball bearings and light bulbs would usually receive a "yes" to this question. So would the base part in an assembly since at least one theoretical part is needed to create an assembly. At the same time, should 25 light bulbs used in an automotive instrument cluster be considered as theoretical parts when only one light bulb shining through a light pipe would work just as well? It is clear that the critical questions of *motion*, *material*, and *assembly/service* must be considered very carefully when deciding if a part is a candidate for elimination. We recommend that an answer of "no" be assigned unless a "yes" is obvious and irrefutable. This maximizes the candidates for elimination and helps stimulate innovative design solutions.

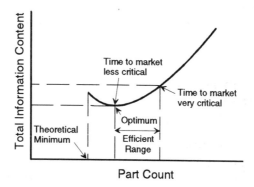

Part Count

Figure 13.1 Achieving an efficient part count depends on a variety of factors. If development time is critical, it may be better to have more parts that are easier to tool and have shorter lead times. If development time is less important, fewer parts, even if they are more complex, will usually result in less total information content. The optimum part count may not correspond to the theoretical minimum, however, if part geometry becomes to complex.

When applying the critical questions, it is also important to remember that the goal in eliminating parts is to minimize overall information content of the design. It is therefore possible that the optimum design might involve more than the theoretical minimum number of parts, especially when the minimum part design requires very complex part geometry (Fig. 13.1). Indications that the part reduction has gone too far include:

- One or more parts are excessively heavy and/or hard to handle.
- Tooling is inordinately costly.
- Lead times are unacceptably long.
- Manufacturing processes are hard to control.
- Manufacturing processes are exceeding best practice limits.
- Designed components are being used in place of standard components.

There are a variety of strategies for reducing part count and part types. We explore some of the most important and effective of these as follows.

13.3 CONSOLIDATE PARTS INTO AN INTEGRAL DESIGN

Integral design, which involves combining two or more parts into one, is possible whenever two or more adjacent parts have been identified as candidates for elimination based on the critical questions discussed above. Also, combining function into one part can often facilitate integral design. For example, the engine block of a high performance motorcycle or racecar can also be used to carry structural load. Similarly, an electronic chassis can be made to act as an electrical ground, a heat sink, and a structural member.

Consolidating parts reduces information content by eliminating separate parts and reducing the amount of interfacing information required. Quality risk and stress concentration due to fasteners and joining processes are also eliminated. Because load paths are smoother and better defined, less material is required and the shape of the part can be adjusted to put material where it is needed. The result is lower weight, increased reliability, and reduced processing information, especially if near net-shape processes such as plastic injection molding, powder metallurgy, and precision casting can be used.

The use of plastic is often a key in integral design. Plastic materials are available for making nearly any part imaginable including springs, bearings, cams, gears, fasteners, hinges, optical elements, and so forth. When used in conjunction with plastic injection molding and other polymer processing methods, these materials can facilitate the consolidation of many parts into one. Some simple examples are illustrated in Fig. 13.2.

A variety of other examples can be observed in modern automobiles and consumer products. For instance, automobile dashboards, including defroster ducts, heater and air conditioning ducts, and so forth, are now routinely molded as one integral part. Similarly, the complex metal tub used in dishwashers, which is built up from many sheet metal parts and must be enameled, has largely been replaced with one-piece plastic injection moldings. A simple one-piece plastic injection molding replaces the complex metal electrical outlet box. A simple, one-piece plastic molding replaces the tile, grout, and caulking of the traditional shower stall. The list of examples is endless.

Along with the many benefits that accrue as the result of eliminating parts, integral design almost always results in a performance improvement as well. The buzz, squeaks, and rattles associated with multipart dashboard construction are eliminated, the possibility of chipping the enamel of the built-up metal dishwasher tub is eliminated, and the potential for water leaks and mildew stained grout is eliminated by the one-piece plastic shower stall design. The improved load paths and lighter weight of integrated parts used in automotive and aerospace applications mean better fuel economy and/or greater payloads.

There are many other materials and manufacturing processes that can be considered when plastic is unacceptable from a functional or cost point of view. For example, powder metallurgy is a good way to eliminate brazed, welded, or staked assemblies of stampings and/or machined parts. Compound gears, cams, links, "multifunction" parts, and other complex parts are also good powder metal candidates. Using extrusions is another highly viable alternative. Tooling

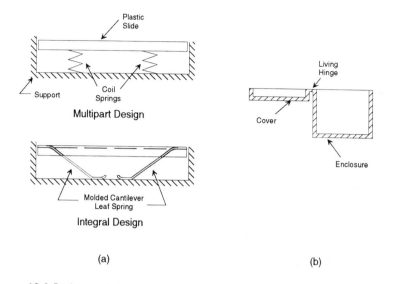

(a)   (b)

Figure 13.2 In integral design, many separate parts are consolidated into one. (a) Two metal coil springs together with nesting, handling, and other assembly problems are eliminated by the molded plastic cantilever leaf springs. (b) Molding the cover and enclosure as one part using a "living" hinge eliminates core pulls, hinge hardware, and alignment problems.

is usually relatively inexpensive, complex transverse geometry is possible, and slide together and/or snap together designs can eliminate many fasteners. Metal casting is another excellent manufacturing process for integral design. The use of thin wall, light metal alloy castings in aerospace and automotive applications can eliminate large numbers of sheet metal parts, rivets, spot welds and the extra material, stress concentration, quality risk, and weight that goes with it.

Integral design involves two important caveats. First, when creating integral designs, great care must be taken to avoid undesirable interactions that may occur due to functional couplings in the integral design. For example, thermal expansion of a combined engine block and structural member may lead to undesirable distortion of the vehicle frame. Secondly, integral designs can lead to complex, "special" parts that are more costly to manufacture than the parts they replace. Usually this added cost is justified however by the cost reduction that results from the elimination of several separate parts. An exception to this can occur, however, when widely used "internal standard" parts are eliminated by the integral design. In these cases, the use of several "internal standard" parts may involve less information content than the design and manufacture of a complex designed part.

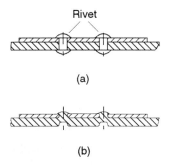

Figure 13.3 Four-piece riveted assembly redesigned as a two-piece assembly by integrating the fastening function into higher level parts. (a) Original riveted design. (b) Redesign using formed features.

13.4 ELIMINATE SEPARATE FASTENERS

Fasteners are always candidates for elimination since they will never receive an answer of "yes" to any of the critical questions. Therefore, integrating fastening functions into higher level parts using the principles of integral design can be an extremely effective way to reduce part counts (Fig. 13.3). In addition, separate fasteners have many undesirable characteristics. In automated operations, fasteners can be very difficult to feed reliably resulting in frequent jams and shutdowns, and they require monitoring for presence and preload. They must be purchased, received, inspected, stored, moved to the point of use, and kept separate to insure that the right fastener goes in the right place. If not properly installed or left lying loose in the assembly, they can present serious quality risk. The information content associated with fasteners is very large. Eliminating fasteners eliminates indirect cost that can be six to ten times the cost of the fastener itself.

The following recommendations provide an effective strategy for eliminating fasteners. Note that these recommendations are ordered according to the amount of information content. When possible, the first recommendation should be followed since it results in the lowest fastening system information content. When this is not possible, then the next possible recommendation should be followed. We refer to a set of design recommendations ordered in this way as *optimal recommendations* since the most optimal design is achieved by implementing the first feasible recommendation for a given design situation.

1. Use "snap together" designs whenever possible.
2. Consider alternative joining processes.
3. Use a minimum number of identical, standard fasteners.

13.4.1 Use Snap-Fits

In addition to eliminating separate fasteners, snap-fits have many superior qualities.

- They are easier to assemble than most other mechanical joining methods. In the assembly process, the snap-fit undergoes an energy exchange, often accompanied by an audible click, that indicates successful assembly.
- The strength of the snap-fit joint comes from mechanical interlocking and friction. It can therefore be designed to be very strong. Also, once assembled, the joint is in a low state of potential energy, which makes it resistant to degradation due to vibration and stress relaxation (creep).
- The snap-in and snap-out force can be tailored to create permanent and non-permanent joints by careful selection of the lead angle and return angle (see Fig. 13.4).

Creating a successful snap-fit design depends on proper consideration of rules governing the shapes, dimensions, materials, and interaction between mating parts. Snap-fit interference is determined by the total deflection of the two mating members. Too much interference can make assembly difficult while too little can result in low snap-out strength. A general design procedure involves the following steps (Chow, 1977):

1. Select the type of snap and the general shape from overall considerations.
2. Determine the spring rate of the spring component.
3. Select interference based on the allowable stress and the elastic limit of the spring material.
4. Specify the lead and return angles to obtain desirable snap-in and snap-out forces.

A snap-fit can fail from permanent deformation or breakage of its spring component, so material selection and environmental factors such as ambient temperature can be important considerations. Snap-fits are also somewhat sensitive to friction and can cease to assemble or disassemble properly if abrasion or oil contamination appreciably changes the friction characteristics of the joint. Dealing with undercut and the associated mold complexity is another challenge in snap-fit design.

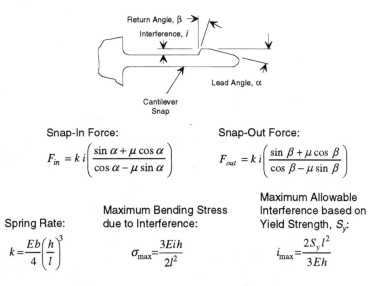

Snap-In Force:

$$F_{in} = k\,i\left(\frac{\sin\alpha + \mu\cos\alpha}{\cos\alpha - \mu\sin\alpha}\right)$$

Snap-Out Force:

$$F_{out} = k\,i\left(\frac{\sin\beta + \mu\cos\beta}{\cos\beta - \mu\sin\beta}\right)$$

Spring Rate:

$$k = \frac{Eb}{4}\left(\frac{h}{l}\right)^3$$

Maximum Bending Stress due to Interference:

$$\sigma_{max} = \frac{3Eih}{2l^2}$$

Maximum Allowable Interference based on Yield Strength, S_y:

$$i_{max} = \frac{2S_y l^2}{3Eh}$$

Figure 13.4 Cantilever snap-fit. Snap-in and snap-out forces are controlled by the selection of the lead angle and return angle, respectively, and depend on the coefficient of friction (μ). Spring rate (k) and bending stress (σ) depend on the width (b) and height (h) of the cantilever cross section, the length of the beam (l), the interference (i), and the material modulus of elasticity (E) and yield strength (S_y) (Chow, 1977).

In some products, repairing or servicing permanent snap-together assemblies can be an issue. One approach is to design the snap so that it can be intentionally broken for disassembly and repair. Reassembly is then performed using self-tapping screws. This approach eliminates the information content associated with fasteners from the product and at the same time, utilizes the advantages of threaded fasteners where they are needed. Although total information content is increased slightly since the service technician must have a supply of self-tapping fasteners, this approach is far superior to using separate fasteners in order to make the few products that eventually require repair easy to disassemble.

13.4.2 Consider Alternative Joining Processes

There are a variety of permanent and nonpermanent joining methods available (Fig. 13.5). Many of these can be considered as possible alternatives for separate fasteners. Each of these joining methods carries with it additional

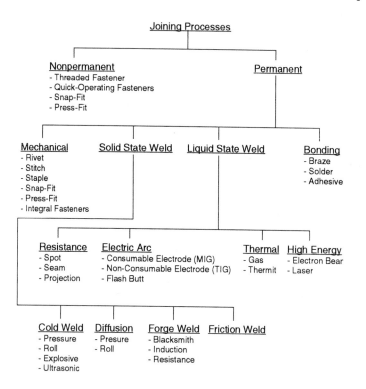

Figure 13.5 Classification of joining processes. Source: Shey (1987).

sources of information content which make them more or less attractive. We briefly discuss some of the alternatives commonly used to eliminate separate fasteners. In-depth discussion of these and other joining methods is provided in most texts on manufacturing processes. See Groover (1996) and Schey (1987), for example.

- *Integral fasteners* involve deformation of component parts so that they interlock to create a mechanically fastened joint. This assembly method requires ductile deformation and is generally limited to sheet metal parts. A variety of approaches are possible. Examples include seaming, in which edges are folded together to form a seam; embossed protrusions, in which bosses are formed in one part and flattened over the joined part (see Fig. 13.3); and stitching, in which lanced protrusions are twisted or bent to interlock with the joined member.

- In a *press-fit* assembly, the mechanical joint is formed by interference fit between the joined components. Press-fits involve simple geometry and do not require undercuts. They are also able to support torsional loading. On the down side, press-fits require close tolerances, introduce residual stress in the joined components, are often performed as a separate operation, and because they are in a high-energy state, they can loosen over time due to stress relaxation and creep.

- *Adhesive bonding* offers simple joint configurations, airtight seals, high strength to weight ratios, compatibility with dissimilar materials, and low stress concentration because of large surface area. Unfortunately, the reliability of adhesively bonded joints depends on surface preparation, curing pressure and time, ambient temperature and humidity, nature of the stress, aging, and other hard to control factors. In addition, special mixing equipment, dispensing techniques and control, holding fixtures, glued joint appearance, spoilage, and shelf life are all concerns. For these reasons, use of adhesive bonding as a means for eliminating separate fasteners could increase information content rather than reduce it.

13.4.3 Use a Minimum Number of Standard Fasteners

By limiting the variety of fasteners used, ordering and inventory problems are reduced, the assembly worker or automation does not have to distinguish between so many separate fasteners, and the variety of fastening tools is reduced. Therefore, when separate fasteners must be used, information content is reduced by limiting the number of sizes and styles of fasteners required. Ideally, only one size and style should be used, and the number of separate fasteners should be held to a minimum. When possible, one-screw assemblies are preferred. In all cases, only standard, commercially available fasteners should be specified. In general, it is best to avoid:

- Screws that are too short or too long.
- Separate washers.
- Tapped holes.
- Round and flat head screws (not good for vacuum pickup).

And, when possible, use:

- Captured washers for reduced part placement risk and improved blow feeding.
- Self-tapping/forming/locking fasteners.
- Screws with dog or cone (chamfered) point for ease of insertion.

- Screw heads designed to reduce "cam-out," bit wear, etc.
- Screw heads having flat vertical sides for vacuum pickup (socket head, fillister head, hex head).
- For blow feeding, shank length to head diameter ratio should be greater than 1:3.

13.5 REDUCE THE NUMBER OF THEORETICAL PARTS

When parts need to be separate because they move relative to each other, finding a design concept that requires the fewest theoretical parts is often the best strategy for eliminating parts. For example, in designing an air compressor, the team might have the option of choosing either reciprocating piston technology or rotating vane technology. Assuming both technologies are functionally acceptable, the rotating vane technology would be preferred because the theoretical minimum number of parts is less for this technology. Obviously the window of opportunity for this strategy is only open for a short time. If the firm is already manufacturing reciprocating piston compressors, the option of using rotating vane technology is less attractive because of business considerations, even though it is preferable from a part count standpoint.

On the other hand, this strategy will often work well, even in mature products, by applying it at the subsystem or component level. This is illustrated by the detent locking mechanism shown in Fig. 13.6. In this example, the old design utilizes tension springs and pawls. In the new design, the detent action is provided by a simple flexure (spring). By using the flexure in place of a complex mechanism, 11 parts are eliminated. Note that the new design (2 parts) is not the minimum part design since the flexure could, theoretically, be integrated into the base plate. This is an example of where the theoretical minimum part design might possibly increase information content.

13.6 CREATE HYBRID PARTS

Hybrid parts are components that combine fabrication and assembly operations (Fig. 13.7). Common examples include insert molding of threaded metal bushings into plastic injection molded parts and joining of components such as shafts and gears using die-casting (Jay, 1971). Such parts are low information alternatives to integral design when materials must be different for reasons of electrical conduction, strength, wear, and so forth. They are also often the least costly alternative for achieving geometry that is too complex or intricate to be incorporated into the mold. Another desirable characteristic of hybrid parts is the very intimate fit that occurs because the cast-on or molded-on material shrink-fits around the insert. Additionally, joint strength is often enhanced because notches on the insert lock mechanically with the enclosing material.

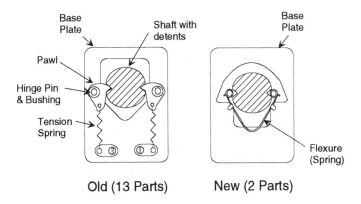

Old (13 Parts)   New (2 Parts)

Figure 13.6 Identify physical concepts that require few parts. In this detent lock example, 11 parts are eliminated by using a flexure in place of spring loaded pawls (adapted from Tipping, 1969).

Hybrid parts can also add information content. For example, the design of the mold or die can become more complicated because features to locate and support the insert must be provided. Also, handling and placing the insert into the cavity is likely to increase the cycle time of the fabrication process. Recycling is another issue since it is more difficult to reclaim the different materials that have been combined to form the part.

Nevertheless, hybrid parts can be an important part reduction strategy in both new product design and in cost reducing existing products. Consider, for example, the vane used in the shutter application shown in Fig. 13.8. In the old design, the vane assembly consists of an extruded aluminum vane and an extruded rubber seal. The rubber seal is assembled into the vane by inserting it into a groove along the vane edge and then crimping the vane to mechanically retain the seal. This assembly process is difficult and time consuming because the seal must be inserted into the vane and crimped manually. In the new design, a structural plastic is substituted for the vane material and a soft, flexible plastic for the seal. The vane is fabricated by co-extruding the two materials together using one die and fabrication process. Information content (and overall cycle time) is reduced tremendously by designing the vane as a hybrid part and combining the fabrication and assembly processes into a single operation.

Another example of the innovation that is possible using hybrid parts is illustrated in Fig. 13.9. In this hybrid part, the electrically conducting terminals are fabricated as a sheet metal stamping and then insert molded as a unit in plastic base. After molding, the connecting piece is cut away to discretize each

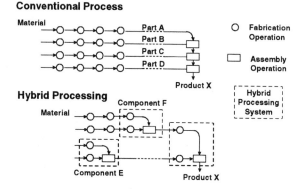

Figure 13.7 Combination of fabrication and assembly operations.

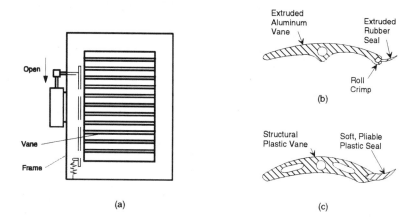

Figure 13.8 Replacing an assembly with a hybrid part: (a) shutter application; (b) cross-sectional view of the original two-piece extruded aluminum vane and extruded rubber seal assembly; (c) cross-sectional view of co-extruded hybrid plastic vane.

terminal. Information content (and cycle time) is reduced significantly by avoiding the need for individual part location features in the mold and the need to handle and place four separate parts within each cycle. A small material cost is incurred since the cut away connection is scrap, but this is small compared to the improvement in production rate and assembly quality (terminals are more precisely and consistently located with respect to each other).

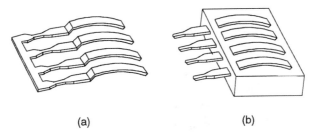

(a)   (b)

Figure 13.9 Reduce information content by postponing descretization. (a) The terminals are fabricated as a sheet metal stamping and assembled as a unit. (b) After insert molding, the connecting piece is cut away.

13.7 STANDARDIZE

One of the most effective ways to eliminate parts is to standardize part designs. In general, there are three strategies for using standardization to eliminate parts.

1. Design so that the same part or component can be used interchangeably in different subassemblies, products, and applications. Parts and components that are designed in this way are sometimes referred to as *repeat parts* or *building block parts*.
2. Create standardized design systems that enable unique new part designs to be easily created, tooled, and introduced into production. Do this by using combinations of standard features, flexible manufacturing systems, special design rules, and so forth.
3. Design new products so that a short list of proven, "off-the-shelf" purchased components can be used everywhere.

Standardization works by eliminating options and reducing the information content of the options that remain. As a result, in addition to producing significant benefits in the form of reduced information content and total cost, standardization can also add cost due to over-design, capital investment, and/or lost sales opportunities. Therefore, benefits and cost must often be traded-off against each other when adopting standardization strategies. For this reason, elimination of parts by standardization is often a business decision as well as a design decision. We discuss standardization further in Chapters 18 through 21.

13.8 KEY TAKEAWAYS

- Elimination of parts is perhaps the single most effective way to eliminate information content from a product design.
- When possible, adjacent parts that are identified as candidates for elimination should be consolidated into one integral part.
- When parts cannot be integrated because of required relative motion, seek alternative design concepts that involve fewer theoretical parts.
- When parts cannot be integrated because materials must be different for fundamental reasons, seek alternative concepts that involve fewer theoretical parts and/or consider hybrid processing.
- Standardization can be a particularly potent strategy for eliminating parts. However, be sure to fully understand the cost/benefit trade-off when considering standardization as a strategy for eliminating parts.
- Choose the part elimination strategy that is most compatible with project requirements relating to product cost, product performance, development time, and development cost.
- When carried to far or inappropriately applied, part reduction can increase rather than decrease information content.
- Some parts have more value than others. For example, a component that is used interchangeably in several products may have more value than a unique part that must be specially designed and tooled for one particular application. Elimination of high value parts should be carefully considered, especially if doing so requires the design of a low value part.

14

Assembly Design

14.1 INTRODUCTION

Assembly design focuses on the development of a coordinated overall part decomposition and detail component geometry that reduces assembly cost by facilitating and easing product assembly. Considering assembly design in the early conceptual stages of design has proven to be extremely effective because, in addition to reducing manufacturing cost, it often generates significant productivity and quality improvements. Assembly design is therefore an exceedingly important consideration, both for redesign of existing products and in new product development. Proper consideration of assembly is even more critical when automated assembly is considered, since cost, cycle time, and complexity of the automation is also directly determined by the product design.

The best way to avoid cost and problems associated with assembly is to avoid the need for assembly. Unfortunately, this is seldom an option because of the many inherent reasons for assembly. For instance, providing for relative movement between parts almost always requires separate parts and therefore assembly. Different materials such as an electrical conductor isolated by an electrical insulator require assembly. Use of purchased components such as a light bulb or hydraulic valve requires assembly. Service modules and replaceable wear parts require assembly.

If assembly can't be avoided, then how should the product be designed to minimize assembly cost? In this chapter, we seek to answer this question by understanding how design decisions drive assembly cost and to use this understanding to develop design strategies that reduce assembly cost and improve assembled product quality. We also briefly discuss formal design for assembly (DFA) methods that can be used to systematically implement these strategies.

14.2 ASSEMBLY COST DRIVERS

Assembly cost drivers are those features of the part decomposition and detail component geometry that determine or establish assembly needs and cost. Analysis of the assembly process shows that, in general, adding a component to the assembly will involve some or all of the following basic functions:

- *Handling*: the process of grasping, transporting, and orienting components.
- *Insertion*: the process of adding components to the work fixture or partially built-up assembly.
- *Securing*: the process of securing components to the work fixture or partially built-up assembly.
- *Adjustment*: the process of using judgement or other decision-making processes to establish the correct relationship between components.
- *Separate Operation*: mechanical and non-mechanical fastening processes involving parts already in place but not secured immediately after insertion (e.g., bending, upsetting, screw tightening, resistance welding, soldering, adhesive bonding, etc.). Also other assembly operations such as manipulating of parts or subassemblies, adding liquids, etc.
- *Checking*: the process of determining that handling, insertion, securing, and adjustment has been carried out properly.

Each of these functions requires time, tools, equipment, and other resources to perform and is therefore a source of assembly cost. Since these functions are generally performed for each part or component that is added to the assembly build, total assembly cost can be estimated as,

$$Assembly\ Cost = \sum_{i=1}^{m} \left(C_H + C_I + C_S + C_A + C_V \right)_i + \sum_{j=1}^{n} \left(C_{SO} + C_V \right)_j \quad (14.1)$$

where m = total number of parts or subassemblies
        n = total number of separate operations
        C_H = handling cost
        C_I = insertion cost
        C_S = securing cost
        C_A = adjustment cost
        C_V = verification cost
        C_{SO} = separate operation cost

Equation (14.1) clearly shows how the product should be designed to reduce assembly cost:

1. Design to reduce the number of parts and the numbers of part types, i.e., reduce the number of separate parts that must be assembled, m.
2. Design to reduce the number of separate operations, n, and to make all separate operations that cannot be eliminated easy to perform.
3. Design parts for easy handling.
4. Design parts for easy mating.
5. Design to eliminate and/or simplify securing processes.
6. Design to eliminate and/or simplify adjustments.
7. Design to eliminate and/or simplify separate operations.

Designing the product in this way reduces assembly cost by reducing information content. Assembly design is therefore essentially an application of guided common sense presented in Chapter 6.

14.3 METHODOLOGY

Assembly design is implemented using the design-analyze-redesign strategy (Fig. 14.1). The process starts by analyzing an existing product or proposed new design for excess assembly information content. Insights gained from the analysis are then used in the redesign phase to reduce the information content of the design. In particular, the team concentrates on reducing information content by eliminating components, separate operations, and adjustments and by designing the parts that remain to be easy to assemble. The methodology is implemented in three steps.

1. **Understand:** Analyze the existing product or proposed new design.
    *   Identify parts that are candidates for elimination.
    *   Identify assembly-related problems and difficulties.
2. **Create:** Develop alternative redesign proposals that reduce information content by eliminating parts, separate operations, adjustments, and so forth and that also simplify and, where possible, standardize the parts that remain.
3. **Refine:** Evaluate the redesign alternatives, and select, improve, and optimize the redesign that best avoids undesirable interactions, has minimal information content, and also satisfies project criteria for:
    *   Product cost
    *   Product performance
    *   Development cost
    *   Development time

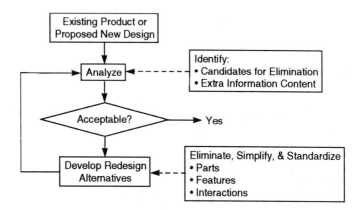

Figure 14.1 Assembly design involves analyzing and redesigning a proposed part decomposition to reduce information content.

Strategies for eliminating parts are discussed in Chapter 13. Therefore, our focus in the remainder of this chapter is on eliminating assembly information content of the parts that remain. Assembly information exists in the form of assembly instructions, handling and insertion motions, randomness of component location and placement, joining processes, potential for assembly mistakes, and effort required correcting a mistake. To reduce assembly information content, components must be easy to handle, insert, and secure without any extra information content due to multiple work surfaces and directions or separate operations and adjustments. Therefore, the basic sources of information content include (1) component handling, (2) component insertion, (3) component securing, (4) assembly direction and assembly surface, (5) adjustments, (6) separate operations, and (7) error checking. *Assembly simplification strategies* involve designing to minimize the information content of each of these sources.

14.4 COMPONENT HANDLING

Component handling is the process of separating a part from bulk, and then grasping, transporting, orienting, and positioning it for placement in the assembly. Factors that effect the ease with which a component is handled and positioned include:

- Component size
- Need for orientation
- Handling impediments and difficulties

Handling information content is increased if components are too small or too large. The following size guidelines apply for ease of manual handling (Boothroyd, Dewhurst, and Knight, 1994):

- The length of the shortest side of the smallest rectangular prism that can enclose the part should be greater than 2 mm.
- For cylindrical parts whose diameter is less than its length, the diameter of the smallest cylinder that can enclose the part should be greater than 4 mm.
- The length of the longest side of the smallest rectangular prism that can enclose the part should be greater than 15 mm.
- Part weight should be such that it is easy to lift and manipulate using one hand without mechanical assistance or special tools.

Similarly, for automatic feeding by conventional vibratory or hopper feeders:
- A part is considered to be too small when its largest dimension is less than 3 mm.
- A part is considered to be too light if the ratio of its weight to the volume of its envelope is less than 1.5 kN/m^3 (0.01 lb/in^3).
- A part is considered to be too large if its largest dimension is greater than 150 mm (6 inches).

The need to orient parts for insertion adds information content. The following guidelines help ensure that parts can be readily oriented either manually or in high speed feeding devices (Boothroyd, Dewhurst, and Knight, 1994).

- When possible, design parts to be as symmetrical as possible (see Fig. 14.2). This will always reduce orientation effort.
- For non-symmetrical parts, emphasize the asymmetry by ensuring that the orientation of the part is defined by one main feature such as an off-center projection, notch, chamfer, or cut-out, which is visible in silhouette (see Fig. 14.3). Avoid parts that are almost symmetrical or that are asymmetrical due to non-geometric features such as differences in surface coatings, lettering, surface finish, and so forth since these require the most effort to orient.
- When possible, design prismatic parts so that the envelope dimensions differ from each other by at least 10 percent. This will facilitate visual orientation in manual assembly and enable the use of simple devices such as wiper blades in automated part feeding systems.

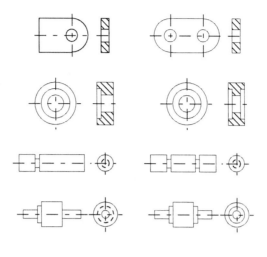

Asymmetrical   Symmetrical

Figure 14.2 Parts made symmetrical for easier orientation and error free assembly.

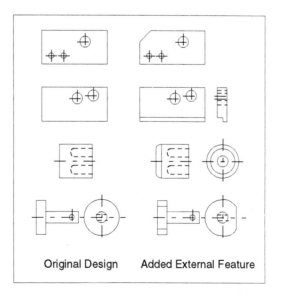

Figure 14.3 Adding external features such as chamfers, slots, and flats can facilitate orientation.

Robotic part handling can be facilitated by providing a large, flat, smooth top surface for vacuum pickup, an inner hole for spearing, or a cylindrical surface or other feature of sufficient length for gripper pickup. Because parts usually come off the production line properly oriented, when possible, this orientation should be preserved by using magazines, tube feeders, or part strips. Palletized trays and kitting are also useful methods for supplying properly oriented parts to the assembly station.

Many part configurations and features can present a variety of handling difficulties. To ensure ease of handling, do the following whenever possible:

- Avoid part features that can mechanically interlock when in bulk by nesting and tangling together. Often a detail design feature such as a narrow gap or curved gap can prevent tangling. Similarly, a varying pitch will prevent coil springs from nesting together. Molded in stops and separators also inhibit nesting.
- Avoid sharp or slippery surface conditions that make parts hard to hold or grasp. Similarly, avoid surface conditions such as grease coatings that cause parts to stick together and/or to stick to the operator's hands.
- Avoid flexible parts that can deform substantially during manipulation or that require two hands to position. If flexure is necessary, then strive to ensure that the parts will retain their shape when handled.
- Avoid fragile or delicate parts that require careful handling or that cannot endure the forces and motions encountered in automated part feeding devices such as bowl feeders and vibratory conveyors.
- Avoid parts that have thin or tapered edges that can overlap or "shingle" when being transported in a conveyor or part-feeding track (Fig. 14.4).
- Avoid sharp corners or edges or other features that present hazards to the operator. When such features are necessary, provide safe handling surfaces and features.
- Avoid gates or other flaws that can interfere with or prevent proper part feeding (Fig. 14.5). Similarly, avoid abrasive surfaces and other features that might wear or damage part-feeding equipment.

14.5 COMPONENT INSERTION

Component insertion is the process of adding components to the work fixture or partially built-up assembly and involves part mating and part placement operations. Factors that effect the ease of part insertion include:

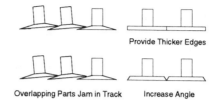

Figure 14.4 Shingling or overlapping can be avoided by providing thicker contact edges or vertical or highly angled surfaces.

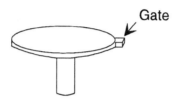

Figure 14.5 Avoid unpredictable inconsistencies such as a gate or other flaw that might act to prevent proper part feeding (Chow, 1978).

- Component quality (dimensional and shape consistency).
- Part features that guide and accommodate assembly.
- Adequate access, clearance, and unrestricted vision.
- Rigidity and accuracy of the base component.
- Selective compliance of the assembly tooling.

The insertion process is often complicated and made more difficult because parts are not always identical and perfectly made. Misalignment and tolerance stack-up produced by dimensional and process variation can produce excessive assembly force. This can lead to increased cycle time, sporadic automation failures, and defective product. High quality components have predictable and controlled variation of features and dimensions that help minimize uncertainty and the risk of jamming and insertion failure. High quality components also reduce precision requirements for part positioning and placement.

Designed-in *guiding features* are part features that guide and ease the insertion process. These include tapers, chamfers, leads, radii, and other features that help align and provide centering force. Generous radii, chamfers, and tapers should be used on all part features that interface and fit together with adjacent parts (Fig. 14.6).

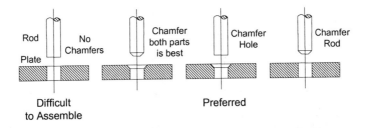

Figure 14.6 Design features that facilitate inserting and mounting of components.

Adequate access and unrestricted vision are essential for ease of component insertion (Fig. 14.7). This design requirement seems so obvious that it is often overlooked. As a result, access problems are often discovered during product launch and are usually very difficult to correct once the product is in production. To avoid access problems, anticipate possible difficulties by planning the assembly sequence early in the design process and imagining how parts will be inserted. Also, be sure to provide adequate clearance for assembly tools and fixtures.

The *base component* is the starting component of an assembly to which other parts are added. A rigid base part helps insure low assembly force because it will not distort or deflect excessively under the action of the insertion force. It also insures that the centering forces developed by the designed-in guiding features act to guide the component being added rather than moving the base to which it is being added. An accurate base part assists assembly automation by minimizing positional variation of assembly features from build to build. If the base part cannot be designed for adequate rigidity, then an appropriate assembly fixture should properly support it.

Selective compliance allows the assembly worker or assembly tooling to correct for component misalignment and placement error. The term "selective" means that the assembly tooling is rigid in the insertion direction so large insertion forces can be developed, but compliant (easy to move) in other directions so it can adjust to accommodate misalignment and placement error. The hands, wrist, and arms of a skilled assembly worker are naturally compliant. The SCARA (Selective Compliance Assembly Robot Arm) robot is rigid in the vertical insertion direction, but is easy to move (compliant) in the horizontal plane. Hence, the guiding forces developed by the designed-in guiding features are able to move the assembly worker's hand or the SCARA robot end effector to align and guide the component during insertion.

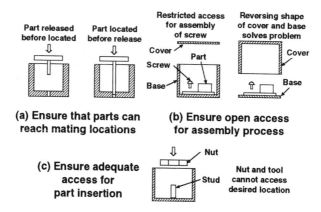

Figure 14.7 Process in the open (Boothroyd, Dewhurst, and Knight, 1994).

High quality components have predictable and controlled variation of features and dimensions. This consistency decreases the range of variation that must be accommodated during the assembly process. Designed-in guiding features, adequate access, selective compliance, and a rigid base, on the other hand, increases the range of variation that can be tolerated.

The above discussion is summarized by the following design for insertion guidelines:

- Provide features such as leads, tapers, chamfers, and generous radii on all mating components. When screws must be used, use dog-points whenever possible. Guiding features reduce assembly automation precision requirements and reduce the risk of jams and misinsertions.
- Ensure that parts, which are not secured immediately upon insertion, are fully located and do not need to be held in place during subsequent insertion operations (Fig. 14.8d).
- Ensure adequate clearance for hands, tools, assembly machine workheads, end of arm tooling, testing, and any subsequent joining processes such as welding or riveting (Fig. 14.7).
- Ensure that the assembly worker can see the surfaces that are to be mated. Avoid performing insertion operations inside enclosures or in regions surrounded by other parts.
- Ensure that the base part is rigid and accurate.

14.6 COMPONENT SECURING

Component securing is the process of physically attaching components to the partially built-up assembly using permanent or non-permanent joining processes. Securing may occur as part of the insertion process (e.g., installation of a threaded fastener) or it may be performed as a separate operation (e.g., adhesive bonding of a joint). A component is designed for easy securing when it is located and retained upon insertion and requires no screwing or plastic deformation as part of the securing operation. Snap-fits, press-fits, circlips, spire nuts, and so forth are examples of components that are easy to secure.

Many securing operations increase assembly information content and should be avoided when possible. For example, if a component needs to be held in place while it is being secured or if it can move or shift position during the securing operation, then extra information content is required making the operation more difficult to perform and the outcome less certain. Similarly, if a joining process requires excessive force, time, or effort, it will, of necessity, involve extra information content. Adhesive joining can require extra time and effort to properly distribute the adhesive and cure the bond. Critical welds must be inspected. Riveting involves plastic deformation and therefor requires special tools and equipment. The integrity of many joining processes is also questionable. For example, the number of spot welds are often increased beyond the minimum required to guard against a hard to detect defective weld.

Design guidelines for minimizing information content associated with component securing are best stated as optimal recommendations:

1. When possible, design so that as components are added to the build, they are (a) correctly located and oriented by features on mating parts (Fig. 14.8*c*), (b) they do not need to be secured immediately, and (c) they do not need to be held in place by an external means. Ideally, the final part should secure all components using a snap-fit.
2. If the component is secured immediately after insertion, avoid screwing operations or plastic deformation of part features. Snap-fits are preferred when possible.
3. If plastic deformation is required, then plastic bending or torsion is preferred to riveting or similar operation.
4. Use dog-point screws to avoid cross-threading. Avoid separate threaded fasteners and washers when possible.

In all cases, joining processes involving uncertainty or excessive time should be avoided. If this is not possible, then the firm should invest in the process to develop it as a core manufacturing expertise.

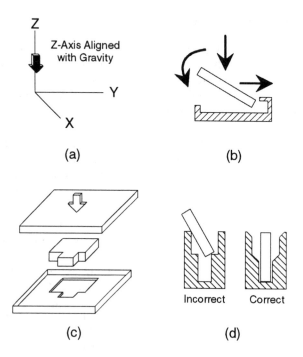

(a)   (b)

(c)   (d)

Figure 14.8 Design for unidirectional assembly: (a) top down z-axis assembly is best; (b) avoid multi-motion insertion; (c) design assemblies as layered stacks with components positively located; and (d) provide alignment features for guiding components.

14.7 ASSEMBLY SURFACES AND DIRECTIONS

The number of assembly processing steps will be decreased by reducing the number of different surfaces to which components are assembled. Ideally, components should only be added to one surface. When more than one surface is involved, the product and assembly sequence should be designed such that all assembly is completed on one surface before moving to the next. Similarly, only one assembly direction should be used. The ideal assembly direction is top down since gravity is assisting rather than opposing the assembly process (Fig. 14.8a). Extra assembly directions mean wasted time and motion as well as more transfer stations, inspection stations, and fixturing. This in turn leads to slower cycle times, increased wear and tear on equipment due to added weight and inertia, and increased reliability and quality risks. The ideal assembly is a layered Z-axis stack with guiding features to align and orient and all components positively located upon assembly (Fig. 14.8).

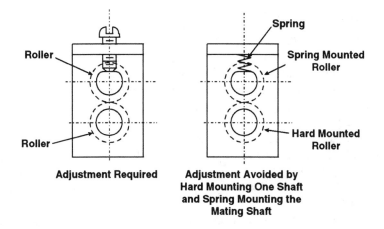

Adjustment Required   Adjustment Avoided by
Hard Mounting One Shaft
and Spring Mounting the
Mating Shaft

Figure 14.9 One way to avoid the need for adjustments is to use compliance to naturally bias the assembly and compensate for misalignment and wear.

14.8 ADJUSTMENTS

Manual and automated mechanical adjustments are expensive and continual sources of assembly, reliability, test, and service problems. Also, equipment that goes out of adjustment is one of the biggest causes of customer dissatisfaction. Avoiding adjustments reduces information content of the design. This translates into reduced assembly cost, higher quality, and easier, less costly, service and maintenance. The need for adjustment can be simplified or avoided in a variety of ways by providing natural stopping points, notches, and spring mounted components, which ensure preferred location as well as compensate for wear (Fig. 14.9). Often, the need for adjustment arises as the result of undesirable interactions (see Chapter 5). In these cases, the need for adjustment can be simplified or eliminated by "decoupling" the undesirable interaction.

14.9 SEPARATE OPERATIONS

Separate operations include all assembly operations other than those directly associated with adding a part, moving to another assembly surface, or performing an adjustment. Examples include mechanical joining processes such as riveting, welding, adhesive bonding, bolt tightening, and so forth. Separate operations should be avoided whenever possible since they add information content in the form of instructions, material handling, floor space, quality risk, to name just a few.

14.10 ERROR CHECKING

Error checking is the process of determining that the handling, insertion, securing, adjustment, and separate operations have been performed correctly. The information content of this process is greatly reduced when the parts are designed so that they cannot be installed incorrectly. Boothroyd, Dewhrst and Knight (1994) suggest the following design guidelines:

- Provide obstructions that will not allow incorrect assembly.
- Make mating features asymmetrical (Fig. 14.8c).
- Make parts symmetrical so that assembly orientation is unimportant (Fig. 14.2).
- Make subsequent assembly impossible when and if a part is incorrectly installed.
- If necessary, provide "clues" such as matching arrows or colors. Note that this is less desirable than ensuring that incorrect assembly is impossible.
- Provide "keys" and other features on flexible parts such as gaskets to prevent incorrect installation. When possible avoid flexible components since these can almost always be incorrectly installed.

Avoid operations that require decision making and judgement during assembly. For example, incorporate critical dimensions into one of the assembled components rather than requiring adjacent components to be carefully adjusted as part of the assembly process.

14.11 FORMAL DESIGN FOR ASSEMBLY METHODS

To be successful, assembly design must be consistently and systematically performed. This has lead to the development of formal design for assembly (DFA) methods that subject all design alternatives to the same evaluation formula. Although sometimes complex and time consuming to use, these methods are helpful because they (1) provide quantitative results for judging ease of assembly, (2) provide the discipline needed to ensure ease of assembly early in the design process, and (3) teach good design practice. Formal methods that have been developed range in complexity from simple handbook approaches to computer software employing artificial intelligence techniques such as expert systems. In addition, many companies and design consultants have developed their own simplified or more specialized versions of DFA. Various design for assembly methods as well as the development and evolution of design for assembly guidelines is discussed in depth by Redford and Chal (1994). Boothroyd, Dewhurst, and Knight (1994) also discuss design for assembly principles and practices in great depth.

14.12 BOOTHROYD-DEWHURST DFA METHOD

As an introduction to formal DFA methods, we briefly describe the Boothroyd-Dewhurst Design for Manual Assembly Method that is widely recognized and used worldwide. The method is presented in Chapters 3 of the book *Product Design for Manufacture and Assembly* by Boothroyd, Dewhurst, and Knight (1994). It is also available in software form from Boothroyd-Dewhurst, Inc., 138 Main Street, Wakefield, RI 02879.

The Boothroyd-Dewhurst DFA method seeks to minimize cost of manual assembly within constraints imposed by other design requirements using a systematic, step-by-step implementation of DFA rules or guidelines such as those discussed above. The process consists of an analysis phase and a redesign phase. In the analysis phase, the time required to handle and insert each component in the assembly is estimated using special charts. Each part is then classified as a theoretical part or candidate for elimination. After each component in the assembly has been analyzed, the design efficiency is calculated which provides a numerical measure of assembly ease.

In the redesign phase, the team uses the analysis results to redesign the product for ease of assembly. The redesign focuses on eliminating and combining parts and on redesigning the parts that remain to have lower handling and insertion times. The new design is then analyzed using the same analysis procedure as that used to analyze the original design. By comparing before and after design efficiencies, the team is able to gage the effect of design changes on ease of assembly.

The charts used to estimate handling and insertion times are based on industrial engineering time study data. The charts are essentially matrices with different times entered in each row-column intersection. Design rules applicable to component handling and insertion are arranged into a hierarchical series of questions on each chart. Estimated handling and insertion times are obtained by answering each question and then branching to different rows and columns based on the answer given. By answering the questions for a particular component, the user is guided to the appropriate row-column intersection containing estimated handling and insertion times for that component. The handling and insertion times determined for each component are summed to obtain the estimated assembly time for each component. Total assembly time for the product or subassembly is calculated by summing all the individual component assembly times.

The design efficiency is defined as the ratio of "ideal" assembly time to "actual" assembly time. The total assembly time for the product or assembly obtained using the charts as described above is used as the estimated *actual* assembly time, T_{actual}. The *ideal* assembly time is estimated as the minimum total time to handle and insert an ideally designed part times the theoretical minimum number of parts, N_{min}. The theoretical minimum number of parts, N_{min}, is determined using the motion, material, and assembly/service questions discussed previously (Section 13.2). Boothroyd and Dewhurst assume that an ideally designed part would require one second to handle, one second to insert, and one second on average to secure for a total assembly time of three seconds per part. The design efficiency is calculated as

$$Design\ Efficiency = \frac{3\,N_{min}}{T_{actual}} \tag{14.2}$$

The basic design for manual assembly procedure is as follows:

1. Obtain the best information about the product or assembly. Useful items include engineering drawings, exploded three-dimensional views, an existing version of the product, or a prototype.
2. Take the assembly apart (or imagine how this might be done). Assign an identification number to each item as it is removed. Initially, treat subassemblies as "parts" and then analyze them as assemblies later.
3. Reassemble the product starting with the part having the highest identification number. As each part is added to the assembly, analyze its ease of handling and insertion and use the three questions to decide if it is a candidate for elimination.
4. When the re-assembly is complete, determine the total estimated assembly time, T_{actual}, the theoretical minimum number of parts, N_{min}, and calculate the design efficiency using Eq. 14.2.
5. Redesign the assembly using the insights gained from the analysis. Analyze the new design by repeating steps 1 through 4 and gage improvements by comparing design efficiencies. Iterate until satisfied.

14.13 KEY TAKEAWAYS

- Because assembly is an integrating process, product assembly can be the source of many manufacturing and quality problems. These problems can often be avoided or mitigated by planning how the product is to be assembled in the early conceptual stages of the product design. Proper assembly design results in low total cost, high productivity, and high manufactured quality.
- In essence, assembly design is an application of the guided common sense strategy developed in Chapter 6 that focuses on eliminating parts and designing the parts that remain to be easy to assemble.

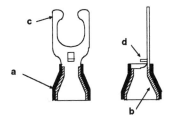

Figure 14.10 Assembly ease is in the details as illustrated by the design of this simple wire terminal. (a) Integral insultation eliminates difficult to install plastic sleeve. (b) A taper guides wire insertion. (c) Tab on tip of prongs provides snap action. (d) A stop helps control insertion depth of the wire (Chow, 1978).

- Assembly design is implemented by following design guidelines that seek to reduce manufacturing cost associated with performing the basic assembly functions of part handling, insertion, securing, adjusting, and checking.
- Correctly implementing the design for assembly guidelines results in (1) inherent ease of manufacture and assembly and (2) designs that are readily automated.
- Designing a product to be easy to assemble requires close attention to design details (Fig. 14.10).
- To be successful, assembly design must be consistently and systematically performed. This has lead to the development of formal design for assembly methods that subject all design alternatives to the same evaluation formula.
- Perhaps the best known and most widely used formal design for assembly method is the Boothroyd-Dewhurst DFA Method.

15

Tolerance Design

15.1 INTRODUCTION

Dividing a product into components and subassemblies is essential for many reasons. Some of these include relative motion between parts, material differentiation, production considerations (no firm is going to manufacture its own light bulb), replacibility and service requirements, and so forth. Unfortunately, once the product is divided into separate parts, how these parts assemble together to create the overall characteristics and features of the product become a key concern. This is because components cannot be perfectly made in large quantities. As a result, key assembly dimensions tend to vary from assembly to assembly due to hard to control dimensional variability of individual components that integrate to establish the overall assembly dimension. This variability in assembly dimensions is commonly referred to as *tolerance stack-up* where tolerance refers to the allowable limits of dimensional variability assigned to each part and stack-up refers to the overall variation in assembly dimension that results.

When not carefully considered and controlled, tolerance stack-up can result in serious functional and manufacturing problems. Tolerance stack-up is therefore a major quality risk for most products. *Tolerance design* is the purposeful planning of the product design to guard against, and if possible, avoid the negative consequences of tolerance stack-up. It involves both the development of the part decomposition and the assignment of tolerances to individual component dimensions. In this chapter, we present a simple methodology for performing tolerance design.

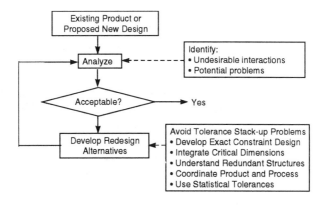

Figure 15.1 Tolerance design involves analyzing and redesigning a proposed part decomposition to guard against, and if possible, avoid the negative consequences of tolerance stack-up.

15.2 METHODOLOGY

Tolerance design involves (1) a systematic search for part decompositions that avoid undesirable tolerance stack-ups, and (2) careful design to mitigate the effects of hard to control dimensional variation when tolerance stack-ups are unavoidable. The methodology is based on the design-analyze-redesign strategy (Fig. 15.1) and consists of the following steps.

1. **Understand:** Analyze the existing product or proposed new design for tolerance stack-up.
   - Identify undesirable interactions.
   - Identify potential functional and manufacturing problems.
2. **Create:** Develop alternative redesign proposals that avoid undesirable tolerance stack-ups where possible and reduce information content of those that are unavoidable.
3. **Refine:** Evaluate the redesign proposals, and select, improve and optimize the alternative that best avoids undesirable interactions, has minimal information content, and also satisfies project criteria for:
   - Product cost
   - Product performance
   - Development cost
   - Development time

15.3 ILLUSTRATIVE EXAMPLE

To illustrate how the methodology is applied, consider the document transport system to be used in a document-feeding device shown in Fig. 15.2. The part decomposition consists of two sheet metal side plates and base plate that form a frame, a sheet metal baffle plate that guides the sheets of paper as they move along the paper path, and rollers that contact the sheets of paper and propel them along the paper path. Bushings, pressed into the sheet metal side plates support the roller shafts. A timing belt and system of pulleys power the rollers.

The drive motor for the paper drive was selected based on drive torque measurements made on a prototype of the product. Subsequently, drive torque for production models has been found to vary significantly from product to product and to be as much as four times the drive torque originally measured in the prototype. It is suspected that alignment of the shaft bearings is responsible. As a fix, tolerances on the sheet metal frame have been tightened significantly and considerable effort has been expended in working with the frame supplier to ensure close conformance to tolerance specifications. In addition, the bushing are now hardened to guard against excessive wear. These measures have not helped appreciably.

The first step in the methodology is to analyze the existing design to identify and characterize the tolerance stack-up problem. Doing this, we identify an undesirable interaction between the component manufacturing process, component joining process, and drive shaft bearing alignment that makes the drive torque highly dependent on tolerance stack-up of the assembled components. The source of the interaction can be attributed to two underlying causes: (1) hole misalignment between the side plates and (2) over-constraint of the baffle plate.

Hole misalignment results from the manufacturing and assembly processes used to construct the frame. The holes for the roller shaft bearing are punched in the sheet metal side plates in the flat. After punching, flanges are formed on the side plates by bending in a brake. The side plates are then spot welded to the frame base to form the frame (Fig. 15.2). Sources of error and variation leading to hole misalignment include:

1. Position and location of the punched holes in the side plates with respect to the formed flanges established by the sheet metal bending process.
2. Angle between the side plate and the formed flanges due to hard-to-control springback in the bending process.
3. Parallelism and distance between the side plates as established by the spot welding process.

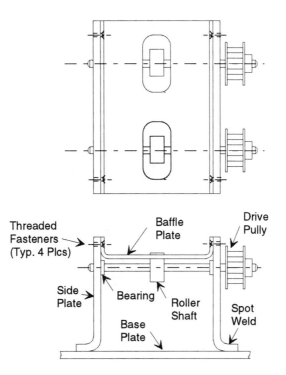

Figure 15.2 Original document transport mechanism design.

Even when tolerances are tightened and dimensional variation of the frame weldment is carefully controlled, the problem is not solved. This is because the width of the baffle plate varies from part-to-part, again because of hard-to-control variation in the sheet metal forming process. Because of *over-constraint* imposed by the method of fastening the baffle plate to the side frames, this variation causes the side plates to bend inward or outward and to warp to varying degrees as the fasteners are tightened. The amount of deformation produced will vary from assembly to assembly depending on the shape and dimensions of the baffle plate relative to the parallelism and spacing of the side plates. This bending deformation will occur to some extent in every assembly unless the shape and dimensions of the baffle plate exactly match the spacing and parallelism of the side plates, which is highly unlikely.

As a result of the undesirable interaction, the drive torque will vary from assembly to assembly according to the degree of deformation induced by the assembly process. The torque measured for the prototype was relatively low because the model maker was careful to insure that the spacing of the side

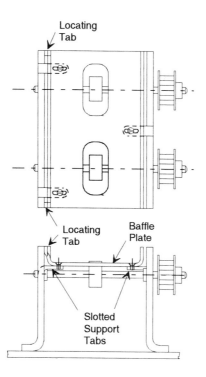

Figure 15.3 Proposed redesign with exact constraint mounting of baffle plate.

plates exactly matched the baffle width. Because exactly matching assembly dimensions is extremely difficult in production, this problem is best corrected by eliminating the undesirable interaction, not by tightening tolerances or working harder to control the manufacturing and assembly processes.

In step 2 of the methodology, we develop redesign proposals to eliminate or correct the problem identified in step 1. One possible redesign for the document transport is shown in Fig. 15.3. This redesign proposal decouples the interaction by reorienting the fasteners so that deflections caused by the fastening forces are orthogonal to the direction that would cause misalignment of the bearings. The fastening system is further decoupled by using slotted or oversize holes so that no transverse forces are accidentally developed during assembly.

A second possible redesign is shown in Fig. 15.4. In this redesign, the undesirable interaction is eliminated by integrating the roller shaft support function and the paper guiding function (baffle plate) together into a single part. The proper alignment of the shaft bushing is then controlled by the

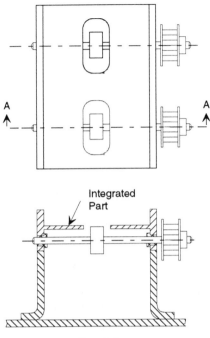

Sec. A-A

Figure 15.4 Proposed redesign with bearing support and baffle plate integrated into one part.

manufacturing process for the integrated part and is no longer dependent on maintaining critical relationships across assembled parts.

Since both redesign proposals (Fig. 15.3 and 15.4) avoid the undesirable interaction, which one should be selected? In step 3, we answer this question by considering information content of the alternative design proposals and the feasibility of each proposal with respect to the project anchors of product cost, product performance, development cost, and development time. In comparing the two redesign proposals, we see that the integrated design of Fig. 15.4 greatly reduces information content by avoiding the need to maintain precise tolerances on the frame assembly. Hence, this redesign proposal should be selected unless development time or cost is crucial, in which case the design of Fig. 15.3 might be more appropriate. Although the integrated part can be designed as a sheet metal part, it should be noted that part-to-part variation of hole alignment might be significantly improved if it was designed as a plastic injection molding or die casting. The choice of manufacturing process obviously depends on the project anchors.

The following important corollary to the principles of good design (Chapter 5) can be drawn from this example:

Integrate functional requirements into a single physical part or solution if they can be independently satisfied in the proposed solution (Yasuhara and Suh, 1980).

The document transport example illustrates how undesirable interactions can be caused and also eliminated by the particular part decomposition selected. Note that the physical concept and the basic functionality of the device are essentially the same for all three designs (Figs. 15.2, 15.3, and 15.4). Note also, that although there is opportunity to eliminate some fasteners, the redesigns did not address design for assembly. Therefore, another iteration to reduce part count and simplify part handling and insertion could further reduce information content of the selected design.

15.4 EXACT CONSTRAINT DESIGN

Over-constraint can be a major cause of tolerance stack-up problems. When a component or subsystem is over-constrained, redundant load paths are created which can result in undesirable deflections and load distributions. These undesirable kinematic constraint interactions cause major structural compromises as well as add cost and complexity to the manufacturing process. *Very often, the need to tighten tolerances is a signal that over-constraint is being masked rather than cured* (Kriegel, 1995). The problem can usually be completely avoided by developing part decompositions that are exactly constrained.

Everything moveable has six degrees of freedom, three translations, and three rotations (Fig. 15.5). The translations are in the x, y, and z directions of three-dimensional space and the rotations are about the x, y, and z directions, respectively. *Over-constraint* of a part occurs when the number of constraints exceeds the number of degrees of freedom.

In *exact constraint* design, each degree of freedom is eliminated by one constraint. The approach is analogous to the 3-2-1 fixturing principle (Fig. 15.5). When a part is set on the x-y plane, it contacts at three points, which removes three degrees of freedom, one translation and two rotations. A second translation and the third rotation are removed by sliding the part along the x-y plane until it contacts the x-z plane at two points. The final translational freedom is then removed by sliding the part along the x-y and x-z planes until it contacts the y-z plane at one point.

In looking at the redesign proposal shown in Fig. 15.3, we see that the baffle plate has been exactly constrained using the 3-2-1 principle. The plate

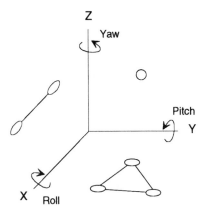

Figure 15.5 Six-degrees of freedom and the 3-2-1 fixturing principle.

mounts on three support tabs, which establishes the position in the z-direction as well as roll and pitch orientation. Sliding the baffle plate to left until the two locating tabs on the baffle plate contact the side plate establishes the position of the plate in the y-direction and the yaw orientation. Finally, the slots on the support tabs, which allow translation in the y-direction but not in the x-direction, establish the position of the plate in the x-direction.

There are many other ways the over-constrained baffle plate in Fig. 15.2 could be more exactly constrained without making major changes to the part decomposition. One simple way would be to replace the fasteners on either the left or right hand side of the baffle plate with index pins. The pins would help locate the plate and provide some redundant support, but distortion of the side frames would be avoided (assuming the baffle plate fits between the side plates) since the plate would not be tensioned by fasteners pulling on each side plate.

Note also, that by following the 3-2-1 rule, the baffle plate in Fig. 15.3 is supported at three points. In many part decompositions, a component or subsystem is supported at four corners. An ordinary kitchen table is an obvious example. In most cases, over-constraint of this kind does not result in major undesirable interactions. At the same time, the prudent design team should be sensitive to potential problems that can arise as a result of this practice. As stated previously, *the need to tighten flatness, parallelity, perpendicularity, and other tolerances of form is a signal that over-constraint is causing an undesirable interaction.*

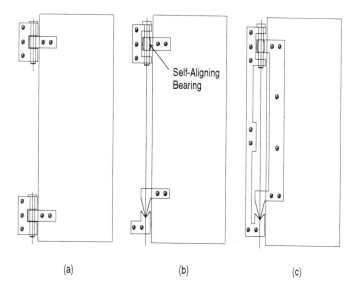

Figure 15.6 (a) Conventional hinged door design, (b) more exactly constrained design, (c) more exactly constrained design with integrated components.

15.5 UNAVOIDABLE TOLERANCE STACK-UP

Tolerance stack-up problems occur when critical features or dimensions must be maintained or controlled during assembly of components. We have seen that many of these problems can be avoided by integrating the critical feature or dimension into a single part or by exactly constraining individual components. Unfortunately, these approaches, *although always preferable*, are not always feasible, especially when components must be separate because they move relative to each other or because they are positioned relatively far apart.

15.5.1 Example of Unavoidable Tolerance Stack-up

As an example of unavoidable tolerance stack-up, consider the brass hinges that are commonly used to support a door and allow it to rotate into open and closed positions (Fig. 15.6a). The hinge pins, which must be collinear to establish the axis about which the door rotates, represent a terribly over-constrained design (Kriegel, 1995). Once one plate of the first hinge is screwed in place, all location information for the other hinge except for height is decided. As a result, the assembly squeaks and binds, and its performance changes with moisture-induced swelling of the door and jambs. In addition, because both hinges can support vertical load, the hinges form redundant supports and the percentage of door weight carried by each hinge depends on fit, alignment, and relative stiffness.

A less over-constrained design is shown in Fig. 15.6b. In this design, the lower hinge is replaced with a pivot point that supports the weight of the door and locates it vertically. The upper hinge is designed so that it supports no vertical force and the self-aligning bearing allows the axis of rotation to adjust so that it passes through the center point of the bearing and the pivot support. As a result, the door is easy to hang (support the door at the pivot, align the upper hinge, and insert the hinge pin), support loads are statically determinant, and the design is robust against hard to control variation such as moisture-induced swelling. Note, however, that information content has been added since the upper and lower hinges are no longer the same and the self-aligning bearing is more complex.

In spite of the functional improvements offered by the design of Fig. 15.6b, the mounting of the door is still subject to tolerance stack-up. For proper operation, the door must fit squarely in the frame when in the closed position, should rotate freely to the fully open position, and must remain stationary at all positions. Because the upper hinge and lower pivot are separate components, this requires that each component be carefully positioned and oriented with respect to the door, the frame, the jamb, and each other. The lower pivot point must be positioned so that the door is properly located with respect to the jamb, and the upper hinge must be positioned to insure that the edge and face of the door are parallel to the jamb along the length of the door. To insure that the door remains stationary in all positions, the pivot and hinge must be positioned relative to each other so that the axis of rotation is vertical within acceptable limits.

One way to help alleviate the tolerance stack-up problem would be to integrate the upper hinge and lower pivot into one assembly as shown in Fig. 15.6c. This helps insure a proper relationship between the hinge and the pivot, but doesn't insure that the axis of rotation will be vertical or that the door is properly located and oriented with respect to the jamb. It also results in a hinge assembly that is overly expensive in terms of volume of material and handling, shipping, and storage.

15.5.2 Use Assembly Fixtures to Control Tolerance Stack-up

A second alternative is to manufacture the door and frame as a pre-hung assembly. In this approach, fixturing is used to locate and hold all components in the correct position and orientation with respect to each other. When all components are properly located and aligned, screws are driven to lock everything in place. For complex tolerance stack-up problems such as this, it can be argued that such an approach may represent the low information content solution. All other solutions add information content, either in the form of extra material, additional and/or more complex parts, and increased handling and storage complexity, or in time, effort and skill required to manufacture and assemble parts to precision tolerances. It is no surprise therefore that pre-hung door assembly with identical upper and lower hinges is widely used in current construction practice.

15.5.3 Avoid Product/Process Interactions

Unfortunately, it can also be argued that, because they can be very complicated, expensive, and time consuming to design, build, operate, and maintain, assembly fixtures are extremely information intensive. For example, the fixture may need to be very precise in order to achieve the tolerances required of the finished assembly. Also, because of rigidity of the fixture, some workpiece tolerances may need to be tight, not because tight tolerance is needed functionally, but because it is needed for the fixture to operate reliably and correctly. Because the design of the fixture usually cannot be started until the design of the product or subassembly is nearly complete, assembly fixtures can entail long and costly lead times. Also, it is not uncommon for undesirable process interactions, such as thermal distortion due to a welding process, to require extensive modification and reworking of the fixture as part of the production ramp-up process.

The key to achieving a low information content product/process solution to unavoidable tolerance stack-up problems is to avoid undesirable interactions between the product and the process. For example, the need to hold tight workpiece tolerances can be avoided by designing the fixture in a way that decouples workpiece tolerance from the finished assembly tolerance. In other words, recognize that components that have considerable dimensional and geometry variation can be combined to form an assembly that is in tolerance. In assembling the door assembly of Fig. 15.6a, it is not important that the upper hinge or lower hinge be made to very precise tolerances, it is only important that they be properly located and aligned with respect to each other before being permanently secured to the door and frame. By employing designed-in tooling features on the components themselves together with appropriate and possibly automated adjustability in the fixture, simple, low information content product design/fixture design solutions can be achieved.

15.5.4 Consider Selective Assembly

A totally different approach might be required in providing a coordinated product/process solution for assembly of precision components where fit is important. To illustrate, consider the tolerance stack-up problem associated with assembling a piston in a bore. Such situations occur frequently in compressor manufacture, internal combustion engine manufacture, and valve manufacture. In these situations, two alternative approaches are available: (1) machine the piston and bore to very precise dimensions on a highly consistent basis from part to part so that all parts fit together interchangeably with acceptable fit or (2) relax the tolerance on the piston and bore dimensions to a level which can be consistently obtained for low cost and selectively assemble by matching the piston size to the bore size. Deciding which of these options is the low information solution obviously depends on trade-offs such as the cost and effort required to make all parts interchangeable verses the cost and complexity associated with measuring, sorting, storing, and matching different size parts. Often, if the fit must be very precise and the cost of customer dissatisfaction is high, selective assembly turns out to be a lower information content solution. This is because the information content associated with measuring and sorting is low compared with that required for *consistent* high precision machining.

15.6 STATISTICAL TOLERANCING

In traditional assembly tolerancing, tolerances are assigned based on a worst case analysis of tolerance stack-up. The *statistical tolerancing* approach, on the other hand, recognizes the stochastic nature of manufacturing. Consequently, tolerances are assigned based on the capability of the manufacturing process and on the probability of failure.

Statistical tolerancing assumes that all dimensions are independent, normally distributed stochastic (random) variables. Each dimension can then be characterized by specifying a mean and standard deviation. To do this, we adopt the following nomenclature:

- The *mean* of a stochastic dimension x is designated as $\bar{x}$
- The *standard deviation* is designated as $\hat{\sigma}_x$

The tolerance specification is related to the manufacturing process by the notion of *natural tolerance*, which is a tolerance equal to plus and minus three process standard deviations from the mean ($\pm 3\hat{\sigma}$). For a normal distribution, this ensures that 99.73 percent of production is within the tolerance limits. Also, since tolerance stack-up involves the summation of dimensions, we note that the sum of two random variables is also a random variable whose mean is the sum of the means of each component and whose standard deviation follows the Pythagorean theorem. For example, the sum z of two random variables x and y is calculated as follows,

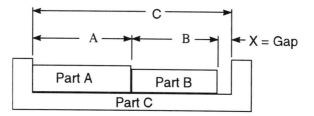

Figure 15.7 Tolerance stack-up example. Note that A = 3.5 ± 0.00105 inches, B = 2.5 ± 0.00105 inches, and $0.0000 < X < 0.0048$ inches.

$$\overline{z} = \overline{x} + \overline{y} \qquad (15.1)$$

and

$$\hat{\sigma}_z = \sqrt{\hat{\sigma}_x^2 + \hat{\sigma}_y^2} \qquad (15.2)$$

To illustrate the statistical tolerancing approach, consider the assembly shown in Fig. 15.7. Suppose that 500 units are to be assembled for the lowest possible cost and that parts A, B, and C are randomly selected for each assembly. The basic dimensions (exact theoretical size) for A and B is 3.5 and 2.5 inches, respectively. The natural bilateral tolerance for both part A (T_A) and part B (T_B) is ±0.00105 inches. The problem of design is to specify a basic length dimension, C, and bilateral tolerance, T_C, for part C. Note that the gap, X, must not exceed 0.0048 inches and that it must not be less than zero if the parts are to assemble without interference.

15.6.1 Worst Case Analysis

In *worst case* analysis, all of the parts are assumed to be at their extreme values. To determine C and T_C so that all parts will assemble without interference and without exceeding a gap of 0.0048 inches, we consider two situations. In the first situation, we assume all of the parts are at their maximum material condition (MMC). In the second situation, we assume all of the parts are at their least material condition (LMC). At MMC, $X = 0$ for the parts to assemble without interference. At LMC, $X = 0.0048$ inches for the gap not to exceed 0.0048 inches. This results in the following equations,

@ MMC, $X = A + T_A + B + T_B - C + T_C = 0$

@ LMC, $X = C + T_C - A + T_A - B + T_B = X_{max}$

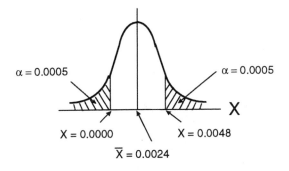

Figure 15.8 Assuming that the gap distribution is centered at $\overline{X} = 0.0024$, 50% of the rejects will be due to interference ($X < 0$) and 50% will be due to the gap exceeding its allowable upper bound ($X > 0.0048$). Since the total probability of a reject must be less than 0.001 for no rejects to occur in 500 trials, $P(X > 0.0048) = P(X < 0) = 0.0005$ and $P(X > 0.0048) + P(X < 0) = 0.001$.

Setting $A = 3.5$, $T_A = 0.00105$, $B = 2.5$, $T_B = 0.00105$ and solving simultaneously gives, $C = 6.0024$ and $T_C = 0.0003$ inches.

15.6.2 Statistical Analysis

As an alternative to worst case analysis, we can select the basic size and tolerance based on the probability that no rejects will occur if parts A, B, and C are selected at random to construct 500 assemblies. To do this, we first calculate the average (basic) dimension for C as follows:

$$\overline{X} = \overline{C} - \overline{A} - \overline{B} = \overline{C} - 3.500 - 2.500 = \frac{0.0048}{2} \quad \rightarrow \quad \overline{C} = 6.0024$$

Now, define P(Rejection) as the probability that the gap will be less than zero or greater than 0.0048 inches (Fig. 15.8). Then, for zero rejects,

$$P(\text{Rejection}) = \left[P(X > 0.0048) + P(X < 0) \right] \times (\textit{Number of Assemblies}) < 0.5$$

since < 0.5 would be rounded to zero (i.e., a half a part can't fail). Therefore, letting n equal the number of assemblies, we have

$$P(\text{Rejection}) = \frac{0.5}{n} = \frac{0.5}{500} = 0.001$$

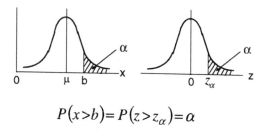

$$P(x>b)= P(z>z_\alpha)= \alpha$$

Figure 15.9 The trasformation variate Z is normally distributed, with mean of zero and a standard deviation of unity.

If we assume that the variation of the gap, X, is normally distributed and that the distribution is centered at $\overline{X} = 0.0024$, then 50% of the rejects will occur because the gap is less than zero and the other 50% will occur because the gap exceeds the maximum allowable gap of 0.0048 inches (Fig. 15.8). Therefore, for no rejects to occur in 500 assemblies,

$$P(X < 0) = P(X > 0.0048) < \frac{0.001}{2} = 0.0005$$

The cumulative density function of the normal distribution is tabulated as a function of the standardized variable Z_α in Table A.1 (Appendix) where

$$Z_\alpha = \left| \frac{b-\mu}{\hat{\sigma}} \right| \qquad (15.3)$$

and b and μ are defined in Fig. 15.9. Linear interpolation in this table gives Z_α = 3.2917 for $\alpha = 0.0005$. Therefore,

$$Z_\alpha^U = \left| \frac{0.0048 - 0.0024}{\hat{\sigma}_X} \right| = \frac{0.0024}{\hat{\sigma}_X} , \quad Z_\alpha^L = \left| \frac{0.0000 - 0.0024}{\hat{\sigma}_X} \right| = \frac{0.0024}{\hat{\sigma}_X}$$

and

$$\hat{\sigma}_X = \frac{0.0024}{Z_\alpha} = \frac{0.0024}{3.2917} = 0.000729$$

From Eq. (15.2), $\hat{\sigma}_X = \sqrt{\sigma_A^2 + \sigma_B^2 + \sigma_C^2}$. Solving for $\hat{\sigma}_C$ gives,

Table 15.1 Cost and manufacturing data for alternative methods of manufacturing part C of Fig. 15.7 (Note: n = number of units delivered)

Alternative Manufacturing Process	Standard Deviation (inches)	Tooling Cost ($/unit)	Machine Cost ($/hr)	Machining Time (min./unit)	Material Cost ($/unit)	Salvage Value ($/unit)
Machine using Manual Mill	0.0010	0.40	15	1.00	2.00	0.25
Machine using CNC Mill	0.0002	0.20	50	1.50	2.00	0.25
Extrude and Cut to Length	0.0008	$\dfrac{1200}{n}$	20	0.25	0.50	0.25

$$\hat{\sigma}_C = \sqrt{\hat{\sigma}_X^2 - \hat{\sigma}_A^2 - \hat{\sigma}_B^2} = \sqrt{(0.000729)^2 - 2(0.00035)^2} = 0.000535$$

Therefore, $T_C = \pm 3\hat{\sigma}_C = 3(0.000535) = 0.0016$ and $C = 6.0024 \pm 0.0016$.

Comparing the worst case tolerance with the statistical tolerance, we see that statistical tolerancing allows us to relax the tolerance on C from ± 0.0003 inches to ± 0.0016 inches. In addition, by using a centered process having a standard deviation of 0.000535 inches or less to produce dimension C, the need to inspect this dimension can be avoided.

15.6.3 Process Selection

Suppose that three alternative methods of manufacture are being considered for producing part C (Table 15.1). Which process should be selected if $C = 6.0024 \pm 0.0016$ inches? One way to answer this question is to estimate the cost associated with each process. This can be done as follows:

$$Cost = Tooling\ Cost + Material\ Cost + Processing\ Cost + Salvage\ Value \quad (15.4)$$

where

   Tooling Cost = (*Units Made*) x (*Tooling Cost/Unit*)

   Processing Cost = ($/hr) x (hr/min) x (min/unit) x (*Units Made*)

   Raw Material Cost = ($/unit) x (*Units Made*)

   Salvage Value = ($/unit) x (*Units Scrapped*)

The number of units made depends on the number of units that are scrapped due to non-conformance with the tolerance specification. Therefore,

$$Units\ Made = Units\ Delivered + Units\ Scrapped \qquad (15.5)$$

and

$$Units\ Scrapped = P(Rejection)\ x\ (Units\ Made) \qquad (15.6)$$

Substituting Eq. (15.6) into Eq. (15.5) and rearranging gives,

$$Units\ Made = \frac{Units\ Delivered}{1 - P(Rejection)} \qquad (15.7)$$

The cost for manually machining part C is estimated by assuming the resulting distribution of manufactured parts is centered (i.e., $\overline{X}_C = 6.0024$ inches) and using Eq. (15.3), (15.4) and (15.7) and Table A.1 as follows:

$$Z_\alpha^U = \left| \frac{C - \overline{C}}{\hat{\sigma}_C} \right| = \left| \frac{6.0040 - 6.0024}{0.001} \right| = \frac{0.0016}{0.001} = 1.6 \quad \rightarrow \quad P(C > 6.0040) = 0.0548$$

$$Z_\alpha^L = \left| \frac{C - \overline{C}}{\hat{\sigma}_C} \right| = \left| \frac{6.0008 - 6.0024}{0.001} \right| = \frac{0.0016}{0.001} = 1.6 \quad \rightarrow \quad P(C < 6.0008) = 0.0548$$

$$\therefore P(Rejection) = P(C < 6.0008) + P(C > 6.0040) = 0.0548 + 0.0548 = 0.1096$$

$$Units\ Made = \frac{Units\ Delivered}{1 - P(Rejection)} = \frac{500}{1 - 0.1096} = 561.54 \approx 562\ \text{Units}$$

Units Scrapped = 562 – 500 = 62 units

$$\therefore Cost = 562 \left[0.40 + \frac{50}{60}(1.00) + 2.00 \right] - 62(0.25) = \$1,473.80$$

$$Cost\ /\ Unit = \frac{Cost}{Units\ Delivered} = \frac{\$1,473.80}{500} = \$2.95\ /\ Unit$$

Table 15.2 Cost of alternative methods for manufacturing part C of Fig. 15.8

Process	Units Made	Units Scrapped	Cost	Unit Cost
Manual Machine	562	62	$1,473.80	$2.95
CNC Machine	500	0	$1,725.00	$3.45
Extrude & Cut	524	24	$1,527.27	$3.11

Performing similar calculations for each alternative manufacturing method gives the results summarized in Table 15.2. Looking at these results, we see that manual machining is the lowest cost option. However, Eq. (15.4) does not include indirect cost. Because of its inherent precision, the CNC machining process eliminates the need for inspection and therefore avoids all of the cost, time, and effort required for inspection and salvage. Avoiding these costs may be well worth the extra $0.50 per part.

15.6.4 Process Capability

Suppose that it is decided to manually machine 500 part Cs and to use these parts to build 500 assemblies without inspecting them for compliance with the $C = 6.0024 \pm 0.0016$ inch specification. If $A = 3.5 \pm 0.00105$ inches, $B = 2.5 \pm 0.00105$ inches, $\overline{C} = 6.00216$ inches, $\hat{\sigma}_C = 0.0010$ inches, and all tolerances are natural, what percentage of the assemblies are likely to be rejected, either because of interference or because the gap exceeds 0.0048 inches? To answer this question, we use Eq. (15.1) and (15.2) to calculate the average gap and standard deviation as follows:

$$\overline{X} = \overline{C} - \overline{A} - \overline{B} = 6.00216 - 3.500 - 2.500 = 0.00216 \text{ inches}$$

$$\hat{\sigma}_X = \sqrt{\hat{\sigma}_A^2 + \hat{\sigma}_B^2 + \hat{\sigma}_C^2} = \sqrt{\left(\frac{0.00105}{3}\right)^2 + \left(\frac{0.00105}{3}\right)^2 + (0.0010)^2} = 0.0011 \text{ inches}$$

Now calculate standardized variables for interference and over size using Eq. (15.3) and use Table A.1 to determine the probability of rejection.

$$Z_\alpha^L = \left|\frac{0.0000 - 0.00216}{0.0011}\right| = 1.96 \quad \rightarrow \quad P(X < 0) = 0.0250 \text{ from Table A.1}$$

$$Z_\alpha^U = \left| \frac{0.0048 - 0.00216}{0.0011} \right| = 2.40 \quad \rightarrow \quad P(X > 0.0048) = 0.0082 \text{ from Table A.1}$$

$$P(Rejection) = P(X > 0.0048) + P(X < 0) = 0.0250 + 0.0082 = 0.0332$$

$\therefore$ Expect about 3.32% rejects or $\approx$ 17 of the 500 assemblies to be rejected.

Comparing this result with Section 15.6.3, we see that only 17 rejects are expected when the lot of part Cs are accepted without inspection compared to 62 rejects if the lot of part Cs is inspected prior to assembly to insure compliance with the $C = 6.0024 \pm 0.0016$ inch specification. This implies that total cost (direct + indirect) will actually be reduced if the parts are not inspected, assuming that rejecting assemblies is no more costly than rejecting individual parts. It further implies that improving process capability (i.e., center the distribution of dimension C and reduce $\hat{\sigma}_c$) will decrease total cost while inspecting parts for conformance increases total cost because it adds information content (inspection and salvage activities) to the manufacturing process.

To test this observation, suppose that the manual milling process used to produce part C is improved by centering it at $C = 6.0024$ inches and reducing $\hat{\sigma}_c$ by 10% to 0.0009. Repeating the above calculations, we find that under these conditions, only 1.95% or approximately 10 of the 500 assemblies would be rejected. If it were possible to reduce $\hat{\sigma}_c$ even further, say by 25% to 0.00075, then only 0.758% or about 1 assembly in 500 would be rejected. This clearly illustrates that improving process capability reduces total cost by eliminating the labor and factory floor space required to perform inspection and salvage activities.

15.6.5 Production Volume

Suppose we needed to manufacture 50,000 part Cs (Fig. 15.7), which of the manufacturing processes proposed in Table 15.1 should be selected? Assuming all distributions are centered and normal, the calculations performed in Sections 15.6.3 and 15.6.4 can be repeated for 50,000 assemblies per year to give the results shown in Table 15.3. Because the cost of the extrusion tool is distributed over the total number of parts delivered, we see that the "extrude and cut" option is the low "direct" cost choice for large volumes.

We also see, however, that, with this high production volume, information content is increased by the number of rejects that can be expected due to the tolerance stack-up. How can the information content due to unavoidable tolerance stack-up be reduced? One possibility is to select the CNC machining

Table 15.3 Expected number of rejects for 50,000 unit production volume

Process for Part C	Part C Inspected Prior to Assembly			Assembly Rejects without Part C Inspection
	Units Made	Units Scrapped	Unit Cost	
Manual Machine	56,155	6,155	$3.60	1,660
CNC Machine	50,000	0	$3.45	0
Extrude and Cut	52,389	2,389	$0.62	539

process to make part C, but this would be extremely expensive from a direct cost point of view. Alternatively, select the "extrude and cut" method to make part C and use selective assembly as discussed previously. Also, endeavor to reduce the variation associated with parts A, B, and C (i.e., $\hat{\sigma}_A$, $\hat{\sigma}_B$, and $\hat{\sigma}_C$) once the product is in production. Unfortunately, it may be difficult or impossible to reduce variation significantly because of resource constraints and/or process limitations.

What can be done "by design" to eliminate this extra information content and totally avoid any direct or indirect cost due to the tolerance stack-up? One possibility is to increase the allowable gap. Recognizing that this may be unacceptable functionally, let us explore the sensitivity of reject rate to maximum allowable gap. To do this, we assume that all distributions are centered at design intent, i.e.,

$$\overline{A} = 3.5000$$
$$\overline{B} = 2.5000$$
$$\overline{C} = \overline{A} + \overline{B} + \overline{X}$$
$$\overline{X} = X_{max}/2$$

where X = Gap. We further assume that the "extrude and cut" process is used to make part C ($\hat{\sigma}_C = 0.0008$ from Table 15.1) and that the tolerances on parts A and B are natural ($\hat{\sigma}_A = \hat{\sigma}_B = 0.00105/3 = 0.00035$). With these assumptions, we can express the standardized variable as a function of maximum gap (X_{max}) as follows:

$$\hat{\sigma}_X = \sqrt{\hat{\sigma}_A^2 + \hat{\sigma}_B^2 + \hat{\sigma}_C^2} = \sqrt{(0.00035)^2 + (0.00035)^2 + (0.0008)^2} = 0.0009407$$

Table 15.4 Expected assembly reject rate due to tolerance stack-up as a function of maximum allowable gap size

X_{max} (inches)	Z_α	P(Rejection)	Expected Number of Rejects in 50,000 Assemblies
0.0048	2.55	0.010800	539
0.0050	2.66	0.007820	391
0.0052	2.76	0.005780	289
0.0054	2.87	0.004100	205
0.0056	2.98	0.002880	144
0.0058	3.08	0.001936	97
0.0060	3.19	0.001374	69
0.0064	3.40	0.000674	34
0.0068	3.61	0.000318	16

$$Z_\alpha = \left| \frac{X_{max} - \overline{X}}{\hat{\sigma}_X} \right| = \left| \frac{X_{max} - X_{max}/2}{0.0009407} \right| = 531.5 \, X_{max}$$

Using this relationship and Table A.1, expected reject rates can be calculated as a function of maximum gap size as tabulated in Table 15.4. This type of data helps provide insight to the design team regarding the trade-off between design decisions and expected assembly rejects due to tolerance stack-up. In looking at Table 15.4, we quickly see that a small increase in maximum allowable gap size will greatly reduce assembly rejects. For example, increasing the maximum allowable gap by 25% (0.0012 inches) reduces the number of expected assembly rejects by about 87% (539 to 69).

15.7 KEY TAKEAWAYS

Tolerance stack-up occurs as the result of dimensional variation of assembled components. Often, many of the problems caused by tolerance stack-up can be avoided by carefully planning the part decomposition and component design early in the design process. This planning process involves understanding how the assembled components of a proposed design interact and using this understanding to iteratively evolve a design that meets cost and performance targets, avoids undesirable interactions, and has the least information content.

- Very often, the need to tighten tolerances is a signal that over-constraint is causing undesirable interactions between assembled parts. These interactions can frequently be decoupled by either (1) employing exact constraint design or (2) integrating critical dimensions into a single part. Generally, the best option will depend on product cost and performance targets and on development cost and development time constraints.

- When tolerance stack-up cannot be avoided, it is often possible to minimize its impact by employing different strategies. One option is to develop a coordinated product and assembly fixture design. In other situations, selective assembly may be the low information content alternative.
- Statistical tolerancing considers the probabilistic nature of tolerance stack-up. Because it is highly unlikely that a worst case tolerance combination will actually occur, the use of statistical tolerancing calculations often allows very tight tolerances to be relaxed somewhat in critical assemblies. It also provides compelling insight into the trade-off that exist between performing inspection, improving process capability, and relaxing tolerance specifications when possible.
- Effective use of statistical tolerancing requires a clear understanding of process capability, including process variation (standard deviation) and the ability to produce to design intent (centered distribution). This in turn requires close cooperation between design and manufacturing.

16

Component Design

16.1 INTRODUCTION

The goal of *component design* is to ensure that designed components are functionally acceptable and also easy to fabricate using the selected material and manufacturing processes. Component design is implemented by creating detail component configurations that minimize information content of the tooling and the process. Tooling information content is minimized when standardized tooling is used and/or the tooling is easily designed and fabricated and has minimal operational complexity. Process information content is minimized when the number of individual processing steps is a minimum, the design specification is well within the process capability, and parameter values are selected to minimize cycle time and cost while also avoiding potential flaws such as porosity, warping, tears, cold solder, and so forth.

16.2 METHODOLOGY

Most manufacturing processes involve a material flow on which shape information is impressed (information flow) and an energy flow which carries out the transformation of information through the tool/die and the pattern of movement for the tool/die and the material (Fig. 16.1). The energy flow includes both the energy necessary to carry out the process and the energy (loss) that is produced by the process. The information flow, which includes both shape and property information, depends on the nature of the material, the type of process (e.g., mechanical, thermal, chemical), the characteristics of the tool/die, and the pattern of movement of the material and the tool/die. The final geometry or shape information of the part is the sum of the initial shape information of the starting material and the shape information impressed on it

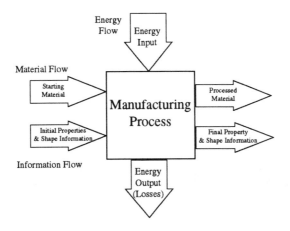

Figure 16.1 Model of a manufacturing process.

by the process. Similarly, the final material properties of the part or product are the result of the property information of the starting material and the property change caused by the process.

For most manufacturing processes, production variables such as tool cost, tool lead time, material cost, process cycle time, and process yield are determined by the interactions that occur between the material and the process. These interactions are heavily influenced and/or controlled by the detail geometry of the component to be produced (Fig. 16.2). Therefore, the design of low cost, functionally acceptable components requires a close matching of component geometry to material and process capability. Specifying component geometry that is compatible with a particular material and manufacturing process usually requires considerable process specific knowledge and experience. For this reason, it is always best to involve tooling and process experts from the beginning of the design process using a team approach. The availability of a broad spectrum of knowledge is especially important in the early stages of the component's design, when the initial configuration and detail geometry are being decided.

These insights form the basis for our approach to component design, which consists of three primary activities:

1. Understand the process limitations and design related requirements.
2. Relate design decisions to component cost and develop insight for making design decisions that reduce component cost.
3. Involve people who deeply understand the manufacturing process early in the design process.

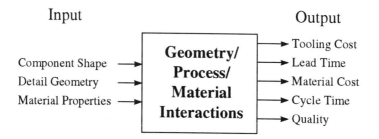

Figure 16.2 Component cost and ease of manufacture depend on how the component geometry interacts with the material and manufacturing process.

Each geometrical feature of a component contributes to both quality and total cost (i.e., direct and indirect cost) of the component. Geometrical features drive tooling cost, labor cost, and equipment cost. They also drive manufacturing activities such as machine setups, material handling, rework, and so forth. The goal of the first activity is to understand process limitations and design related process requirements in terms of component geometry, quality, and total cost and to use this understanding to develop fundamental guidelines for good component design.

The goal of the second activity is to relate geometry decisions to cost. As a first step in doing this, a detail cost model is developed that includes both initial investment and recurring costs. Geometry decisions that drive these costs are then identified and analyzed to provide insight into how component geometry drives total component cost. Finally, it is often useful and/or helpful to express this insight in terms of design rules and guidelines that can be used to help guide the design of components that have low total manufacturing cost.

Ultimately, good component design is the result of having all needed design information available early in the design process when initial geometry decisions are being made. The goal of the third activity is to ensure the availability of deep process expertise when it is needed. Experience has shown that even the most well-intentioned designer, working by him- or herself, usually doesn't have the detailed knowledge required. The only sure way to avoid unneeded design iterations and the time and cost penalties that result is to involve process experts from the start.

In the sections that follow, we illustrate the application of this methodology to a variety of component design situations. We then discuss how a structured team approach combined with computer simulation and other design techniques can be used to help improve component design practices.

16.3 PROCESS-SPECIFIC DESIGN

Process-specific component design has to do with the design of parts to be manufactured using particular manufacturing methods or processes such as casting, forging, injection molding, and sheet metal stamping. A large variety of different manufacturing processes are widely used to produce individual piece parts (Fig. 16.3). It is interesting to note that most of the these manufacturing processes as well as many others can be synthesized using the rational building block method discussed in Chapter 9 by employing different combinations of material flow, energy flow, and information flow building blocks (Fig. 16.4). For more in depth discussion of individual manufacturing processes, the reader is referred to the many excellent textbooks on manufacturing processes and methods that are available. See, for example, Groover (1996) and Shey (1987).

For a given component, the particular manufacturing process selected depends on the part geometry that is required and on the material that is to be used. General selection criteria include consideration of (1) functionality, (2) production volume, and (3) project constraints (e.g., product cost, product performance, development cost, and development time). The component design approach works well for most process-specific design situations as illustrated by the following examples.

16.3.1 Design for Plastic Injection Molding

To illustrate the approach, consider plastic injection molding, which is a near net-shape manufacturing process that is widely used to produce large quantities of production parts. In plastic injection molding, plastic material is heated to form a "melt" which is then forced under pressure into the mold cavity to form the part. Because the material is molten when injected into the mold, complex shapes and good dimensional accuracy can be achieved. Molds with moving cores and unscrewing mandrels allow the molding of parts with undercuts and internal and external threads. The molds may have multiple cavities so that more than one part can be made in one cycle. Proper mold design and control of material flow in the mold cavities are important factors in the quality of the product. Other factors affecting quality are injection pressure, temperature, and condition of the resin.

Process Limitations and Requirements

There are three major product/process interactions that must be considered in the design of a plastic injection molded part: material shrinkage, gating location, and parting line selection. Shrinkage occurs as the molten plastic cools in the mold to form the solid part. Since thick sections cool more slowly

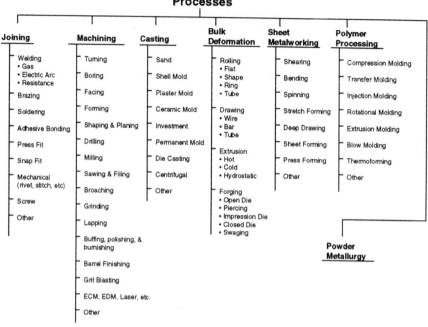

Figure 16.3 Classification of commonly employed manufacturing processes.

Material Flow		Energy Flow		Information Flow	
State of Material	Basic Process	Energy	Medium of Transfer	Pattern of Movement for Material	Pattern of Movement for Tool/Die
Solid	Mechanical	Mechanical	Rigid	No Movement	No Movement
Liquid	Thermal	Electrical	Elastic	Translation	Translation
Granular	Chemical	Magnetic	Plastic	Rotation	Rotation
Vaporous		Thermal	Liquid	Translation/ Rotation	Translation/ Rotation
		Chemical	Granular		
			Gaseous		

Figure 16.4 Rational building block matrix for generating alternative manufacturing processes.

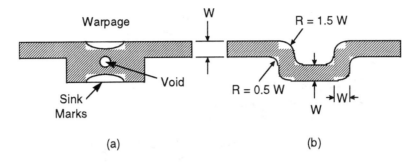

Figure 16.5 (a) Shrinkage defects occur due to variations in thermal mass in different regions of a plastic injection molded part. (b) A uniform wall thickness (nominal wall) combined with generous fillets and radii help avoid shrinkage problems.

than thin sections, most shrinkage defects occur when different regions of the part have different section thickness (Fig. 16.5a). In such cases, warpage and shape distortion can occur as the part cools because thinner sections cool more quickly and shrink less than thicker sections. Also, because "cooler" regions are more rigid than "hotter" regions, stress can develop due to constraints that develop as the part contracts onto the mold cores (male parts of the mold). Often, these stresses are "frozen-in" as the part solidifies resulting in undesirable residual stress that degrades the parts strength and functionality.

Shrinkage defects are eliminated or avoided by employing a uniform wall thickness everywhere in the part (Fig. 16.5b). The uniform wall thickness is often referred to as the *nominal wall* since it is the target value when no other wall thickness is specified. By using a nominal wall, thermal mass, and therefore shrinkage, is the same everywhere in the part. Avoiding sharp corners also reduces shrinkage problems (Fig. 16.5b). Sharp corners, especially inside corners, cause severe molded-in stresses as the material shrinks onto the core. Sharp corners also cause poor flow patterns, reduced mechanical properties, increased tool wear, and stress concentration (Muccio, 1991).

A well-designed plastic injection molding can typically be broken down into three basic elements: the nominal wall, projections off that wall, and depressions in that wall (Beall, 1985). Once the basic configuration and functional features of the part have been decided, nominal wall design is implemented by isolating each element and designing it to have a uniform wall thickness (Fig. 16.6). By decomposing the overall part into basic elements, it is relatively easy to correctly design the part as a plastic injection molding, no matter how complex it might be.

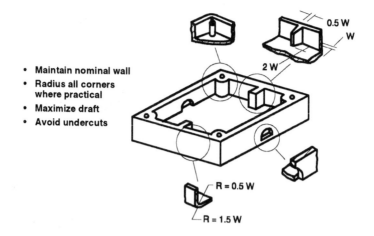

- Maintain nominal wall
- Radius all corners where practical
- Maximize draft
- Avoid undercuts

Figure 16.6 Isolate each element in the design and consider it individually (adapted from Beall, 1985).

The *gate* is the connection between the runner system and the molded part. In general, the location of the gate as well as its type and size, strongly affect the molding process and the physical properties of the molded part. Often, the location and design of the gate must be considered during the design of the part. For example, the gate location can be critical if defects and/or possible part failure due to the weakening effect of weld lines is to be avoided (Fig. 16.7). Similarly, the gate location is critical when the orientation of glass fibers or other filler is important. Also, the gate location can be important from an appearance or assembly standpoint (see Fig. 14.5). Typically, the location and dimensions of the gate are based on experience or on analysis using commercially available computer programs. Once the mold has been fabricated, its performance is tested and, if necessary, the gate geometry is refined by "trial and error."

When a mold closes, the core and cavity, or two cavities meet, producing an air space into which the plastic is injected. From the inside, the mating junction between the mold halves appears as a line. This line also appears on the part and is called the *parting line*. A part may have several parting lines. The selection of the parting line is largely influenced by the shape of the part, tapers, method of ejection, type of mold, esthetic considerations, post-molding operations, inserts, venting, wall thickness, number of cavities, and the location and type of gating. Selection of the parting line is a very important decision because it impacts many aspects of the part. Consider for example a part having a general rectangular shape. It is seen that the parting line determines:

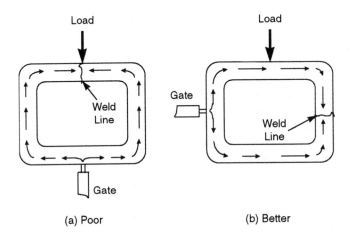

(a) Poor   (b) Better

Figure 16.7 Weld lines form when the plastic flow divides into two or more paths. (a) This gate location causes a weld line to form in a critical region of the part. (b) This gate location causes the weld line to form in a less critical region.

- The detail shape of the part.
- The surfaces that must have draft.
- The number of undercuts.
- The surfaces on which the ejector pins can act.
- Where and how the part is gated and the mold vented.
- The tolerances that can be maintained.
- The projected area of the part, which in turn determines the clamping force required during the molding process.

To facilitate proper release of the injection molded part from the tool, part surfaces that are perpendicular to the parting line must be tapered in the direction of mold movement. This taper, which is commonly referred to as *draft*, allows the molded part to break free by creating a clearance as soon as the mold starts to open. Draft is necessary to offset the effect of shrinkage, which causes the plastic part to grip cores making part ejection from the mold difficult. Although there are exceptions, a draft of $1/2°$ per side is considered a minimum, with $1\ 1/2°$ to $3°$ per side frequently recommended. In general, the larger the draft, the easier the part will be to eject. This in turn will allow faster cycle times since the part does not need to cool as much in the mold before it is strong enough to be ejected.

Undercuts are geometrical features that prevent mold opening and mechanical ejection of the part. Side action (camming) is used to overcome the effect of an undercut. Molding of internal threads and some types of external threads require unscrewing devices. Camming and unscrewing devices add considerable complexity and cost to the mold and should be avoided whenever possible. One trade-off is to avoid the cost of camming by utilizing secondary operations such as drilling or machining. This can be undesirable however because it replaces a one time tooling cost with a recurrent production cost. A more preferable approach is to eliminate camming by designing the part and selecting the parting line such that functional and appearance requirements are satisfied without undercuts.

Design Related Cost Interactions

The cost of a plastic injection molding is determined by its size, configuration, and detail geometry. By understanding how geometry decisions effect the cost, it is possible to make geometry decisions that reduce cost. Developing this understanding is a three-step process:

1. Develop a cost model based on the major sources of recurring and non-recurring cost associated with the part.
2. Identify the key design related cost drivers.
3. Understand the relationship between the design related cost drivers and the cost sources. Develop insight into the effect of each cost driver on cost and how part cost can be reduced.

Step 1 is implemented by first identifying the major cost sources and then expressing part cost in terms of these cost sources. For plastic injection molding, the major sources of cost include tool cost, material cost, and production cost. Unit part cost is calculated as follows,

$$Cost / Part = \frac{C_T}{N} + V C_M + \frac{C_H t_{cycle}}{Y} \qquad (16.1)$$

where C_T = total tooling cost ($)

C_M = material cost ($/in^3)

C_H = machine cost including labor ($/hr)

N = lifetime number of parts made using the tool

V = part volume (in^3)

t_{cycle} = cycle time to mold one part (hr)

Y = process yield (usable parts/N)

Equation (16.1) is very clear as to what should be done to reduce cost:

- Design to minimize tooling cost.
- Design to minimize material cost.
- Design to minimize cycle time.
- Design to maximize process yield.

In step 2, we use our understanding of the manufacturing process to identify the design decisions that drive these costs. For plastic injection molding, the major design related cost drivers include the parting line, the number and complexity of undercuts, the nominal wall thickness, the amount of draft, surface finish specifications, and dimensional tolerance specifications.

Once the key design related cost drivers have been identified, the relationship between each driver and part cost is delineated in step 3. For plastic injection molding, these relationships are summarized as follows:

Nominal Wall Thickness: Of all the issues in plastic injection molding design, selecting the proper nominal wall thickness is probably the most important and all-encompassing design decision that is made. Just about every aspect of the parts appearance, functionality, and manufacturability relates to, affects or will be influenced by the wall thickness. An in-depth discussion of all factors that must be considered in selecting the nominal wall is beyond the scope of this example. From a cost standpoint, however, it is safe to say that the nominal wall thickness will impact all of the cost sources in one way or another. The most obvious is material cost since wall thickness directly effects the part volume with a thinner wall being more desirable. Also, a thinner wall will facilitate faster cooling and as a result, potentially shorter cycle times. At the same time, a thinner wall requires more injection pressure and time to fill the mold. If the wall is to thin, yield may be adversely effected because flow is restricted and the part will not fill out properly. Also, high injection pressure can result in flash or highly stressed parts. In addition, thinner walls can require more mold detail and closer core/cavity alignment and fit which will increase tooling cost. Thick-walled parts, on the other hand, cool slower, shrink more, and have more risk for sink marks and voids. Thick walls also create more random plastic flow patterns, which reduce the dimensional stability and predictability of the part. Hence, in addition to increasing material cost, thicker walls can also decrease yield and increase cycle time.

Parting Line: The parting line establishes many aspects of both the part geometry and the process. The parting line choice effects the size of the injection molding machine required and the complexity and size of the mold.

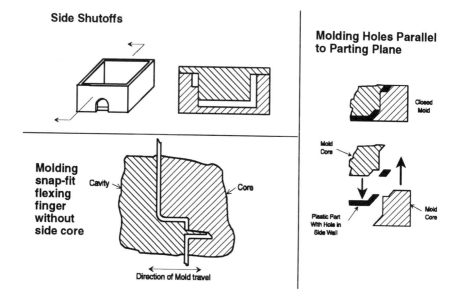

Figure 16.8 Creating "undercuts" with simple tooling.

This in turn effects tooling cost, machine cost, and to a lesser extent cycle time. To minimize part cost, the parting line should be selected very early in the design process. The goal should be a parting line that keeps tool complexity to a minimum. A planar parting line is preferable to a non-planer parting line. Also, when possible, a parting line that places the whole part in one mold half can greatly reduce tool cost since only one mold half needs to be machined and alignment and registration requirements across the parting line are avoided.

Undercuts: Anything that can be done to avoid undercuts or reduce their complexity will reduce tooling cost and development time. Also, since complex side actions require additional time, avoiding or simplifying undercuts can often reduce cycle time. The best way to avoid undercuts is to select the parting line wisely. In addition, it is often possible to design so undercuts can be created using simple tooling without the need for expensive side actions (Fig. 16.8).

Draft: Generous draft is highly desirable for low part cost. Minimal draft, especially when combined with deep draws (Fig. 16.9), increase tool cost because highly polished mold surfaces or special mold surface treatments are required for proper part ejection. Alternatively, a regular application of mold

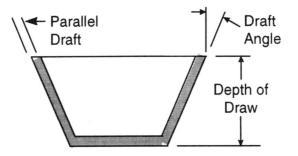

Figure 16.9 Generous draft enables reliable part ejection and short cycle time.

release spray may be necessary during production. This increases cycle time and production cost. Minimal draft also increases cycle time because the part must be stronger and therefore cooler before it can be reliably ejected. Yield may also be adversely effected when difficult part ejection results in damaged parts. A difficulty that is commonly encountered when applying draft to a part is the creation of unacceptably heavy walls. One remedy is to use parallel draft which allows the wall sections to be kept uniform (Fig. 16.9).

Surface Finish: The appearance and feel of the finished part depends on the surface finish. Textured surfaces and very smooth surfaces increase tool cost because of the extra effort required to prepare the mold surface. Surface finish can also adversely impact cycle time and yield if it is not properly designed. To avoid problems, care must be taken to ensure that the draft of the part is adequate to allow proper part ejection. Also, the texture features must have positive draft.

Tolerance: An advantage of the plastic injection molding process is that it allows good dimensional control. However, tight dimensional and form tolerance specifications increase tooling cost and reduce cycle time and yield. Therefore, whenever possible, component tolerances should be relaxed as much as possible.

The relationship between the cost sources and design related cost drivers for plastic injection molding is illustrated graphically in Fig. 16.10. Simple graphic depictions such as this can be a helpful aid for creating cost consciousness. Many of the relationships and design insights discussed above can be reduced to design guidelines and rules. We illustrate this in the design for machining example, which is the topic of the next section.

Cost Drivers \ Cost Sources	Parting Line	Under-cuts	Nominal Wall	Draft	Surface Finish	Toler-ance
Tooling Cost	●	●	◖	◖	●	◖
Material Cost	○	○	●	○	○	○
Cycle Time	◖	◖	●	●	◖	●
Yield	○	○	◖	◖	●	●

● Major Influence   ◖ Moderate Influence   ○ Little Influence

Figure 16.10 Relationship between design related cost drivers and cost sources for the plastic injection molding process.

16.3.2 Design for Machining

Machining is a material removal process in which the part is generated by moving the workpiece and the tool relative to each other using a variety of machine and tool combinations (see Fig. 16.3). Because the shape information is impressed by tool motion rather than tool geometry, the machining process has the great advantage of being flexible and easily adapted to a variety of part sizes, shapes, and detail geometry. Machining is therefore often the process of choice for low volume production. In addition, the machining process is capable of generating very complex and highly precise geometry. For this reason, it is often used as a secondary process to complete parts that are initially formed using a primary process such as casting or welding.

Process Limitations and Requirements

There are three major process characteristics that must be considered in the design of parts that are to be machined. First, the process must be setup. This involves mounting the workpiece in the appropriate work holding fixture and adjusting the machine to perform the desired machining operations. Setup takes time so what ever can be done to simplify or avoid setup is a plus. Another limitation is the relationship that exists between surface finish, cutting conditions (i.e., depth of cut, cutting speed, and feed rate), and tool life. From a cycle time standpoint, it is desirable to machine as fast as possible. However, this can result in unacceptable surface finish and excessive tool cost and tool change time. Hence it is usually necessary to use cutting conditions that effectively balance these factors. Finally, machining processes often involve considerable handling and manipulation of the workpiece as well as moving the workpiece from machine to machine. As in setups, anything that can be done "by design" to eliminate and minimize non-value added material handling is a plus with respect to cost.

Design-Related Cost Interactions

The primary cost sources for machining include machine cost, setup cost, tool cost, and material cost. Cost per part can be calculated in terms of these costs using the following cost model:

$$
Cost = \sum_{j=1}^{m} \sum_{i=1}^{n} \left[\left(\frac{C_0 \left(t_{np} + t_m + t_{ch} \left(\frac{t_m}{T} \right) \right)}{Y} + C_t \left(\frac{t_m}{T} \right) \right)_i \right]_j + \frac{C_{Mat'l}}{\prod Y_{ij}} + \sum_{j=1}^{m} \left(\frac{C_0 t_{su}}{N} \right)_j
$$

(16.2)

where m = the number of machines and/or setups
        n = the number of operations for each machine and/or setup
        C_0 = machine and operator cost ($/min.)
        C_t = tool cost ($)
        t_{np} = non-productive time (min.)
        t_m = machining time (min.)
        t_{ch} = tool change time (min.)
        T = tool life (min.)
        Y = yield for each operation (1 − scrap rate)
        $C_{Mat'l}$ = material cost ($/workpiece)
        t_{su} = setup time (min.)
        N = number of parts in batch

Analysis of Eq. (16.2) provides the design guidance needed.
- Design to minimize machining time
- Design to minimize machine cost
- Design to minimize tool cost per part
- Design to minimize non-productive time
- Design to maximize process yield
- Design to minimize cost of raw stock

Cost Sources

Design-Related Cost Drivers	t_m	t_{np}	C_0	C_t	Y	$C_{mat'l}$
Volume of Material Removed	●	○	◐	◐	○	●
Material Properties	●	○	◐	◐	○	●
Tolerances	●	◐	●	◐	●	○
Surface Roughness	●	○	○	○	◐	○
Number of Operations	○	◐	●	●	●	○
Number of Setups and/or Machines	○	●	●	◐	●	○
Part Size	●	◐	●	◐	◐	○
Number of Non-Standard Features	◐	◐	○	●	○	○
Tool Clearance	○	◐	○	●	○	○
Access to Surface to be Machined	○	●	○	◐	○	○
Length of Tool Path	●	○	○	◐	○	○

● Major Influence   ◐ Moderate Influence   ○ Minor Influence

Figure 16.11 Relationship between design-related cost drivers and cost sources for machining processes.

- Design part to suit available raw material sizes
- Use near net shape preform (cored holes, etc)
- Divide into multiple parts if appropriate

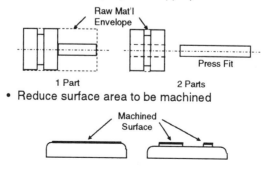

- Reduce surface area to be machined

Figure 16.12 Example ways to reduce the volume of material removed.

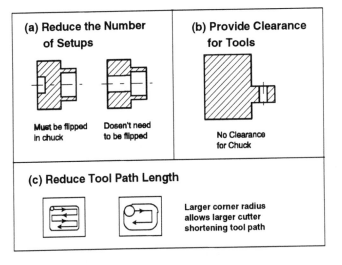

Figure 16.13 Approaches for simplifying the design of a machined part.

Design Guidelines

Following a process similar to that illustrated for plastic injection molding, a variety of design related cost drivers are identified and correlated with cost sources as shown in Fig. 16.11. Often, the relationship between the design related cost drives and cost sources can be reduced to a set of design guidelines for low cost component design. Using the insights provided by Fig. 16.11, we list the following set of design for machining guidelines:

- Reduce the volume of material to be removed (Fig. 16.12).
- When possible, use low cost free machining materials.
- Avoid tight tolerances.
- Avoid excessively smooth surfaces.
- Drive toward a minimum number of operations.
- Drive toward a minimum number of machines and setups, e.g., design so that all features can be machined from one direction (Fig. 16.13a).
- When possible, reduce part size to allow the use of smaller, less costly machines and material handling equipment.
- Provide clearance for tools (Fig. 16.13b).
- Insure access to surfaces.
- Specify features that allow shorter tool path lengths (Fig. 16.13c)
- Use a rationalized set of standard features; avoid non-standard features whenever possible (Fig. 16.14).

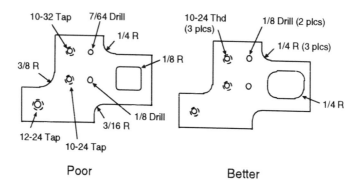
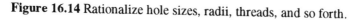

Poor   Better

Figure 16.14 Rationalize hole sizes, radii, threads, and so forth.

16.4 FACILITY-SPECIFIC DESIGN

Facility-specific design refers to the design of components and subassemblies that are to be made using a highly specialized or unique manufacturing facility. Typical examples of such facilities might include a special purpose laser-cutting machine for sheet metal or a specialized facility for fabricating hybrid microelectronic circuits. A facility-specific design guideline typically expresses the capability limitations of the facility as design rules. For example, the design rules for the laser cutting facility might specify:

- Allowable material types and thickness.
- Maximum workpiece envelope size.
- Minimum allowable slot width.
- Minimum spacing between holes.

Similarly, design rules pertaining to the microelectronic fabrication facility would likely specify allowable geometry and configurations such as the following:

- Minimum spacing between bond pads.
- Minimum die to bond spacing.
- Minimum obstruction and bond pad spacing.
- Minimum metalization trace width.
- Minimum spacing between two metalizations.
- Minimum component spacing.
- Allowable component orientations.

In addition to specific design rules, a facility-specific design guideline might also include a variety of other design information and aids. For example, various physical models and examples illustrating desirable as well as undesirable practices may be provided. Also, specialized information about the facility, expressed in a readily usable and relevant form for the designer might be included.

To illustrate a facility-specific application, suppose a manufacturing firm has invested in a flexible assembly system (FAS) that can be programmed to automatically assemble a variety of different subassemblies used in various products sold by the company. Such a system would likely consist of several stations intended to perform certain classes of assembly operations. Individual parts might be handled and inserted by robots using programmable "grippers" and fixtured using programmable fixtures especially designed for the FAS. For a particular subassembly to be built on a specific FAS facility of this kind, the subassembly and parts making up the subassembly must be correctly designed to be compatible with the available assembly operations, the available work envelopes, and the associated flexible tooling that interfaces the parts and in-process assembly to each station and to the material handling system. The design guideline for such an FAS would provide needed process specific information to ensure conformance of the product design with the FAS requirements. It would also facilitate "what-if" optimization and experimentation with alternative design concepts and configurations.

Facility-specific design guidelines are often essential if specialized manufacturing systems are to deliver the productivity and quality improvements that they promise. Most importantly, development of an effective and user-friendly guideline usually depends on close cooperation and coordination between those who are most familiar with the facility and those who design the components that are to be fabricated or processed by the facility.

16.5 STRUCTURED TEAM APPROACH

Component design is an iterative process (Fig. 16.15a). The problem of design is typically formulated in terms of functional requirements and constraints that must be satisfied. Functional requirements relate to the functions the part must provide while constraints relate to the form (shape, size, surface finish, precision, etc.) and processing (parting line, draft, section thickness, etc.) requirements that constrain the geometry that can be selected. Based on the problem formulation, an initial design is created. This design is then evaluated and modified iteratively until an acceptable design is achieved. Typically, the redesign is guided by the design information, insight, and understanding developed in the evaluation step. To be acceptable, the design must satisfy all functional requirements and constraints.

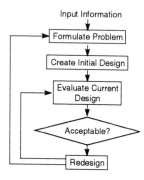

(a) Iterative design process.

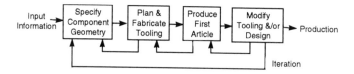

(b) Traditional component design.

Figure 16.15 If not carefully structured, component design can involve a multitude of design iterations.

Similar iterative design processes are performed at each stage of the design. Hence, when the traditional approach to component design is examined, we see that a multitude of iterations can occur during the component development (Fig. 16.15b). First the design engineer goes through the iterative design process to specify the component geometry. This geometry is then passed on to the manufacturing or tooling engineer who repeats the iterative design process to specify the tool and process design. Problems discovered during this stage generate additional iterations if component geometry changes are required. Additional iterations to the component geometry and tool design may also be required during tool fabrication and preparation for production of the first article. Finally, iterative changes to the tooling and perhaps the component geometry may be necessary to "tweak" the design to meet production requirements.

Excessive design iterations significantly increase design cost and time. Most importantly, design iterations performed late in the process can lead to suboptimal design. The result is a component that falls short of cost and performance targets. Such designs place the whole project in jeopardy.

How can the traditional component design process be improved? We propose several guidelines based on the above discussion.

1. Design the component geometry and process as a coordinated system in one concurrent process. Consider geometry, material, and process interactions and design related cost drivers from the beginning as part of the process.
2. Develop a thorough understanding of all customer needs including downstream processing constraints before beginning the component design.
3. Focus on creating an acceptable initial design. By spending the time "up front" to create the best possible initial design, a large number of lengthy analyze-redesign iterations are avoided. The evaluation phase should confirm the design rather than create it.
4. Use manufacturing process simulation software and other modern computer-aided analysis and inspection methods to quickly optimize the design.
5. Develop a consistent, well-defined "science base" for component design by developing design guidelines and structured methodologies for each core manufacturing process or method used by the firm.

The goal of these suggestions is to shorten the component design cycle and help ensure that the best possible component design is created. In looking at the list of suggestions, we see that suggestion 5 is essentially implemented by activities 1 and 2 of the general component design approach presented in Section 16.2. Implementation of suggestions 1 through 3, and to some extent, suggestion 4, on the other hand, requires multidisciplinary input. As has been discussed extensively elsewhere in this book, multidisciplinary input is best implemented by utilizing a structured team approach.

The goal of the team approach is to have all required product and process knowledge available when the key early design decisions are being made. In a *structured team approach*, the overall problem of design is broken down into a series of sequential, easier to perform steps that proceed from the general to the specific. Excessive iteration and long design times are avoided by performing each step in a thorough and disciplined manner. In general, each step in the process can be further subdivided into steps to create a hierarchy of structured methodologies.

To illustrate the structured team approach, we propose a simple meeting based component design process (Fig. 16.16). The use of meetings as a means for effective communication and input is predicated on the recognition that not

Figure 16.16. Simple "meeting-based" structured team approach.

all team members can be available for designing the component on a continuous basis. Team meetings are therefore scheduled at which all salient aspects of the design are reviewed and discussed. All members of the team must be present at these meetings. The purpose of the meetings is to establish design direction, make key design decisions that require input and consensus from all team members, make sure that all process constraints and requirements are being properly considered, and resolve conflicts and impediments to the proposed design. The outcome of each meeting is a set of action steps to be implemented by individual team members. In this way, all team members are kept informed and participate in the design decision making process. At the same time, the actual detail work of creating the design is delegated to specific team members according to the skills and knowledge required. In the following, we briefly discuss each step of the methodology and imagine how a component would be designed using this methodology.

Step 1: Form Team

This is the pivotal first step in the methodology. Unless the arrangement is formalized in some way, it often is difficult to get effective collaboration between the design and manufacturing engineer early in the component design process, especially before the component configuration and geometry are defined. By being formally assigned to a team, each individual team member takes personal responsibility for the design from the beginning. This fosters and facilitates the kind of collaborative attitude that is essential for good design.

All "stakeholders" who have an interest in the component should be represented on the team. A typical team might include a design engineer, a process engineer, a tooling design engineer, and perhaps one or more specialists who are familiar with process simulation software, finite element analysis, fracture mechanics, non-destructive evaluation (NDE) techniques, and so forth. In addition, if the component is to be purchased from an outside supplier, it is essential that the supplier be properly represented on the team. This not only helps ensure that all customer and processing needs are appropriately considered, it also makes it possible to rapidly negotiate changes to the design specification when necessary, and to quickly assess cost consequences of design decisions.

Meeting 1: Clarify the Design Problem

Clarifying the problem consists of developing a general understanding of the cost, performance, and manufacturing goals and constraints of the design. A typical agenda for this meeting might include the following:

1. Review of product background.
2. Customer requirements and design objectives.
3. Expected annual production volumes and target costs and lead times.
4. Geometry concepts and alternatives.
5. Material and processing options.
6. Production facility and secondary processing locations.
7. Potential geometry/material/process interactions.
8. Developing a preliminary configuration design.
9. Making assignments to team members to create the initial design.

As mentioned above, it is essential that all team members be present at each team meeting. For example, although the analyst may not be actively involved with the design before the component geometry is fully defined, it is extremely important that he or she participate in the early design decisions that lead to the proposed geometry. In this way, the analyst knows all the needs of the design problem and is familiar with the reasoning behind the particular geometry that will eventually be analyzed and optimized.

Step 2: Create the Initial Design

The initial design establishes the detail layout of the component geometry and the tooling and/or process concept. It includes the configuration and parametric design of the part together with tooling and processing information such as parting line selection, gating location, tooling surfaces, datums, and so forth. *Configuration design* involves the determination of what features such as walls, holes, ribs, etc. will be present and how these features will be connected to provide the desired form, fit, and function. *Parametric design* involves the determination of dimensions, tolerances, and exact material specifications needed to meet durability, stiffness, natural frequency, and other functional and performance targets.

As a general approach, a preliminary configuration might be proposed in Meeting 1 by the team as a whole. Using this as a starting point, the design engineer and manufacturing engineer work together to develop the details of the configuration design, seeking input and consensus from various team members as necessary. Once the initial component configuration has been firmed up, a preliminary parametric design would be performed. The goal of this task is to quickly determine section dimensions, secondary processing requirements, and material property requirements using simple strength of materials methods or, if necessary, a rough finite element analysis. Once the approximate parametric design is complete, the overall design is evaluated and modified to minimize cost. This is easy and straightforward to perform with a minimum of analysis and iteration because the component geometry, secondary processing, and tooling concept have been conceived and developed as a coordinated system with input from all team members. Hence, all of the information needed to make quality design decisions is readily available.

Meeting 2: Refine and Approve the Initial Design

The goal of this meeting is to react as a team to the initial design and to make any adjustments or modifications deemed necessary by general consensus of the team. This is the time when all design and processing issues should be discussed and resolved. If there are significant impediments to the design as proposed, these should be resolved before proceeding to Step 3. A typical agenda for this meeting might include the following:

1. Review the initial design.
2. Identify impediments, potential undesirable interactions and performance and processing concerns.
3. Discuss all design-related costs to ensure that the best component geometry from a total cost standpoint has been identified.
4. Make assignments to team members to work out solutions to various impediments and schedule a follow on meeting.
5. If no impediments are identified, approve the initial design and make assignments to team members to refine and optimize the design.

Step 3: Refine and Optimize the Design

Once the team is confident that the initial component geometry and tooling design is the best solution possible, the effort required to optimize details of the component geometry and tooling concept by computer analysis can be justified. The goal of this step is therefore to computer model the design and iteratively improve it until all aspects have been appropriately optimized. The amount of effort expended on this step will depend on how important it is to optimize the component. For example, if weight and/or material cost is critical, extensive effort to minimize the amount of material used can be justified. Similarly, if safety is an important issue, comprehensive analysis to ensure acceptable fatigue life and reliable detection of flaws can be justified. The key to this step is to start with a component geometry that is close to the optimum. This will minimize the time, analysis effort, and number of iterations required to converge to the optimum design.

Meeting 3: Approve the Final Design

The result of Step 3 will be a fully specified component design including the detailed component geometry, tooling design, and fixture design. In addition, the finished component design including machining, heat treating, and so forth will be fully specified. The purpose of Meeting 3 is to formally review the finished design as a team and approve the design for release to manufacturing. When the structured team approach is performed properly, the final design will almost always be approved. However, if the team decides that the design is not ready to be released, then appropriate action plans for correcting design deficiencies must be developed and implemented. One or more follow-on meetings may then be required before the design is released.

Although painful at times, not releasing the part until it is ready helps insure a minimum number of tooling changes and "tweaks" and, in the long run, is the most cost effective policy. By strictly adhering to this policy, the component design should proceed quickly and smoothly to first article and production with little or no modification. When this is the case, the team knows that it has done its job well.

16.6 KEY TAKEAWAYS

The goals of good component design are (1) a rapid and efficient design process, (2) a functionally acceptable part that can be manufactured for the lowest possible total cost and highest possible quality, and (3) a design that transitions quickly and smoothly into full scale production. Achieving these goals requires the following:

- A clear understanding of all manufacturing process requirements and limitations.
- Design guidelines that relate design decisions to total cost and provide the insight needed to achieve high quality and performance for minimum cost.
- A design process that effectively utilizes people who have deep expertise in the manufacturing process to be used in producing the component.

17

Manufacturability Improvement Method

17.1 INTRODUCTION

The *Manufacturability Improvement Method* is a systematic approach for analyzing an existing or proposed product design to identify opportunities for simplifying and improving efficiency of manufacture and assembly. The method seeks to minimize information content by eliminating parts and processes, simplifying the parts and processes that remain, and standardizing where possible. An "analyze", "create", and "refine" procedure is used. In the *analysis* phase, the product is disassembled and then systematically reassembled. As each part is added to the build, it is analyzed for ease of manufacture and assembly. In the *create* phase, insights gained from the analysis are used to develop several redesign concepts and/or ideas for improved ease of manufacture and assembly. These concepts are then evaluated, winnowed, combined, and optimized in the *refine* phase.

17.2 METHODOLOGY

The primary focus of the methodology is on reduction of information content by using guided common sense (see Chapter 6) to eliminate parts and separate operations and to simplify the parts that remain by designing for ease of manufacture and assembly. Part count reduction is especially important because the information content of a manufactured product is directly proportional to part count. Eliminating parts is therefore the quickest and most effective way to reduce total cost. Similarly, ease of assembly is emphasized because assembly is often the single most important source of indirect cost and quality risk. When productivity and quality improvements are sought, design for ease of assembly must therefore be given the highest priority. Creating an easy to assemble design is even more important when automated assembly is considered, since cost, cycle-time, and complexity of the automation is directly determined by the product design.

The starting point for finding creative ways to eliminate parts is to question the need for each part in the proposed design. This is done by asking the basic "motion", "material", and "assembly/service" questions discussed in Chapter 13 (Section 13.4). If the part receives a "no" answer to all three questions, it is a candidate for elimination (CFE). One of the goals of the redesign phase is to find creative ways for eliminating CFEs. In seeking to eliminate parts, it is important, however, to note that some parts have more value than others. For example, the team should carefully weigh the consequences of eliminating a standard part if doing so requires that a new unique part must be designed.

Ease of assembly is evaluated by analyzing each part for extra information content required to perform the following assembly functions:

- *Handling*: the process of grasping, transporting, and orienting components.
- *Insertion*: the process of adding components to the work fixture or partially built-up assembly.
- *Securing*: the process of securing components to the work fixture or partially built-up assembly. Securing may occur as part of the insertion process (e.g., installation of a threaded fastener) or it may be performed as a separate operation (e.g., heat stake or ultrasonic weld).
- *Adjustment*: the process of using judgement or other decision-making processes to establish the correct relationship between components.
- *Separate Operations*: mechanical and non-mechanical fastening processes and other assembly operations involving parts already in place.
- *Checking*: the process of determining that the assembly process has been performed correctly.

In the redesign phase, creative approaches are then sought to eliminate extra information content by simplifying component and assembly design to ease and facilitate the performance of these functions.

17.2.1 Design Guidelines

The manufacturability improvement method is, in essence, a systematic application of the cardinal design rules discussed in Chapters 5 and 6.

- Avoid undesirable interactions.
- Minimize information content.

Among the designs that avoid undesirable interactions, the best design is the one that has the minimum information content. Information content is reduced by implementing the following design guidelines.

Optimize the Assembly Sequence. Develop an assembly structure that clearly delineates the assembly sequence of components as well as the standardization and manufacturing strategy for the product. Consider a "building block" approach, or a layered or stacked construction, or the use of a base component that locates, orients, and relates the various components of the product to each other. Select a starting point for assembly that facilitates a minimum number of assembly operations. The starting point may be a base component, a subassembly or module, or some other portion of the assembly envelope. Using the starting point selected, visualize the order of incorporation of parts into the assembly envelope as well as the spatial relationships involved and seek to minimize the information (e.g., the number of individual instructions) required to carry out the assembly. Anticipate where interference between previously assembled parts and subsequent operations might occur and design to simplify or eliminate potential trouble spots.

Design for a Minimum Number of Parts. Always seek the minimum number of simply shaped components. Fewer parts mean less of everything that is needed to manufacture a product. A part that is eliminated costs nothing to design, change, make, assemble, move, handle, purchase, inventory, rework, repair, or replace. Consolidate parts by combining mating or contacting parts that do not move independently of each other into an integral part. Consider alternative materials, net shape manufacturing processes, and so forth.

Eliminate Separate Fasteners. Eliminate fasteners by integrating their function into mating parts. Fasteners themselves may be relatively inexpensive but their installation is costly and ripe with quality risk. If all fasteners cannot be eliminated, design to reduce the fastener count. Consider "snap together" or interlocking designs. Rationalize standard fasteners to reduce fastener types, head styles, drive styles, sizes, etc.

Eliminate Adjustments. Identify critical dimensions that, if not confined to a single part, require slots and other features to permit adjustment between parts. If possible, incorporate such dimensions into a single part. Avoiding the need for adjustment between mating parts reduces assembly cost, simplifies automation, improves product reliability and durability, and eliminates quality risk. Equipment that goes out of adjustment is a major cause of customer dissatisfaction.

Eliminate Randomness. Use rigid parts wherever possible. Flexible parts are more difficult to handle and orient. Plan the layout of electrical connections, flexible tubing, and control cables early in the design. Provide features that orient, locate, and restrain wire harnesses and other components that would otherwise occupy random orientations or positions. Avoid dangling connectors, lead wires, unrestrained parts, ambiguities between mating parts. Avoid part to part or part to process dependencies such as "fit at assembly".

Design Parts for Easy Handling. Design symmetrical parts. Part symmetry means fewer orientation steps, resulting in faster assembly time and lower automated handling system costs. If symmetry is difficult or asymmetry is important, then design easily recognized orientation and positioning features or provide clear identifying marks. When possible, avoid very large or very small sizes. Try to use rigid parts and avoid flexible components. Provide adequate, easily recognized and accessed surfaces to allow mechanical gripping for transfer and location retention. Design to eliminate part tangling, nesting, and interlocking.

Reduce Processing Surfaces. Simplify components so that the number of surfaces being processed is minimized. Try to design parts for orthogonal and/or parallel fabrication since most fabrication equipment is designed using orthogonal and/or parallel constructs. Try to complete all processing on one surface before moving to the next. This is especially important in assembly. Don't design the assembly sequence so that the operator must jump back and forth from surface to surface.

Minimize Insertion Motion. Avoid complex, multi-motion insertion. Develop a "top-down Z-axis" assembly approach. Z-axis assembly means simple robot or insertion tooling with gravity serving to assist in the assembly process. Tooling is usually less expensive; part clamping becomes either not necessary or less costly. Try to design so that parts are assembled using a "sandwich" approach with unidirectional Z-axis insertion. Not using Z-axis assembly usually means special tooling for both fixtures and automatic equipment.

Design Parts for Easy Insertion. Design for easy part insertion by providing generous tapers, leads, chamfers, and radii on mating parts. Such guiding features help align and provide centering forces to assist and facilitate the insertion process. They also offset the detrimental effect of part inconsistencies, misalignments, and tolerance stack-up.

Process in the Open. Design out restricted access and obstructed vision. Provide adequate clearance for *standard* tooling. Provide a clear, unobstructed view for assembly operations. This is important for eye-hand coordination for manual assembly and, in most cases, essential for automatic assembly.

Retain Parts. Once parts are in place, keep them in place. Design location features that eliminate re-positioning or holding down a component after it has been "mated" to another part. Provide these features either in the part being assembled or in its mating part.

Design for Error Free Assembly. Design parts so that they are impossible to assemble incorrectly. Provide tabs, "nests", and location features that prevent incorrect mating of parts. Poor quality is no longer an issue since it is impossible to achieve.

Reduce, Simplify, and Group Processes. Use low technology on the factory floor. Design to reduce the number of different processes and operations required for product manufacture and assembly and to allow the use of simple mechanized processes. Develop a short list of preferred processes and practices and use these in new designs. Avoid processes that require specialized or unusual operator skills or training. Avoid difficult to control processes. Group processes for more efficient line operation and increased opportunities for automation.

17.2.2 Step-By-Step Procedure

The method utilizes an analyze-redesign approach to implement the design guidelines. First, an existing design is analyzed. The insights gained from the analysis are then used to develop and refine redesign alternatives aimed at eliminating parts and making the parts that remain easy to manufacture and assemble. The step-by-step procedure is as follows:

1. **Gather Information:** Obtain the best information available about the product or assembly. Useful items include:
   - Engineering drawings
   - Exploded three-dimensional views
   - An existing version of the product
   - A prototype

2. **Analyze:** Determine design improvement opportunities.
   - Take the assembly apart (or imagine how this might be done). Treat subassemblies as "parts" and then analyze them later as assemblies.

- Begin reassembling the product in the reverse order from which it was disassembled. Imagine adding each part in the Z-axis "top down" direction using one hand. Reorient the build if necessary to make this possible. In doing this, imagine a special fixture that holds the partially built-up assembly together and allows the build to be oriented so that the next part can be added in the "top down" direction.
- Complete a line of the *Manufacturability Analysis Worksheet* (Fig. 17.1) for each part as it is added to the assembly. Similarly, complete a line on the worksheet for each reorientation, adjustment, or other separate operation (e.g., bending, riveting, bonding, etc.).
- Analyze each part as it is added to the build. Analyze first for ease of handling, insertion, securing, and checking. Then determine if the part is a candidate for elimination (CFE) using the "motion", "material", and "assembly/service" questions. A part that receives a "no" to all three questions becomes a CFE. Note that all separate fasteners, reorientations, adjustments, and separate operations are, by default, CFEs.
- Determine the value of the part or operation. Internal standardized parts (building block parts) and external standard parts (purchased parts) have high value, while a new part that must be designed has low value. Separate fasteners, reorientation, and adjustments have low value. Separate operations that are considered "core" expertise and are widely used would have greater value than operations that are seldom used or are hard to control or require special skills and equipment.
- Rate the manufacturability of the part. A part that is simple and is properly designed for the economically best manufacturing process would receive a high rating, whereas a part that is complex or is difficult to economically manufacture would be rated low.
- Review the design for *undesirable interactions*.

3. **Create:** Redesign the product to take advantage of opportunities and insights identified in Step 2. The goal is to (1) eliminate parts and separate operations, (2) eliminate all negatives (-'s) of the parts that remain, and (3) maximize the number of pluses (+'s). To increase the possibility of identifying a "best" redesign option, develop several alternative concepts ranging from simple (and practical) changes to speculative (and risky) but desirable "radical" ideas.

4. **Winnow, Refine, Optimize:** Develop improved concepts by eliminating negative features and combining positive features. Evolve a final, best design using the principles and techniques discussed throughout this book.

1	2	3	Assembly				Part Eliminaton				Assessment			
			4	5	6	7	8				9	10	11	12
Part or Operation	Qty	Type	H	I	S	C	Motion	Mat'l	Ass'y	CFE	V	M	UI	Notes

$\sum Qty$ = $\sum CFE$ =

Figure 17.1 Manufacturability analysis worksheet.

17.2.3 Column-By-Column Worksheet Guideline

1. **Part or Operation Name.** Enter a brief name for the part or operation.

2. **Quantity:** Enter the number of parts or the number of times the operation is performed.

3. **Type:** Identify the type of part or operation as follows:

    0   Separate operation (e.g., reorientation, adjustment, etc.)
    1   Separate fastener
    2   Separate component
    3   Subassembly, to be treated as a part now and analyzed later

4. **Part Handling (H):** Rate handling of the part as follows (leave blank for separate operations):

Rating	Condition
+	Part has been designed for easy handling • Part is easy to grasp and manipulate with one hand • Orientation is easily achieved - part is symmetrical or has a readily apparent asymmetry
0	Part presents no handling difficulties
-	Part is difficult to handle • Part is difficult to orient • Part tends to nest, tangle, stick together • Part is delicate, flexible, has sharp corners and/or edges • Part requires grasping aid or two hands

5. **Part Insertion (I):** Rate insertion of the part as follows (leave blank for separate operations):

Rating	Condition
+	Part has been designed for easy insertion • Accessible - part and tools can reach the desired location • No holding down is required • Easy to align and position - features provide position and orientation information • No resistance - generous radii and guiding surfaces are provided
0	Part presents no insertion difficulties
-	Part is difficult to insert • Access and/or vision is obstructed • Part is unstable after insertion - holding down is required • There is appreciable resistance to insertion - part can jam, wedge, hang-up, tight clearance

6. **Securing (S):** Rate the securing operation as follows (leave blank for separate operations):

Rating	Condition
+	Part is designed for easy securing • Part is located and retained upon insertion • No screwing operation or plastic deformation required (e.g., snap-fit, press-fit, circlip, spire nut, etc.)
0	Securing operation presents no difficulties • No securing required • Easy to align and position with no resistance • Required force, time, and/or effort are reasonable • High inherent quality
-	Securing operation presents difficulty • Alignment and/or position is difficult to achieve and/or maintain • Required force, time, and/or effort is excessive • Process is unreliable or uncertain

7. **Checking (C):** Rate checking of the assembly operation as follows:

Rating	Condition
+	Assembly operation cannot be performed incorrectly • Fixture nest provides orientation and prevents error • Obstruction prevents incorrect assembly • Mating features are asymmetrical • Orientation is not important (e.g., part is symmetrical) • Parts have been specially marked (e.g., matching arrows or colors, etc.)
0	Operation requires no special checking
-	Operation must be checked • Alignment and/or position must be checked • Strength or soundness must be checked • Process parameters must be monitored and/or controlled (e.g., bolt torque, curing time, liquid quantity, etc.)

8. **Candidate for Elimination (CFE):** Place the quantity entered in column 2 in the CFE column for separate fasteners and separate operations such as adjustments and reorientations, which are candidates for elimination by default. A part receiving an answer of "NO" to **all three** of the following questions is a *candidate for elimination* (CFE). A part receiving an answer of "YES" to **any** of the questions is a *theoretical part* and usually cannot be eliminated for fundamental reasons of motion, material or assembly.

1) Does the part move relative to all other parts assembled?
2) Must the part be made of a different material from the other parts already assembled because of fundamental reasons such as electrical conductivity or insulation?
3) Must the part be separate for reasons of assembly or disassembly?

Place the appropriate answer to each question ("YES" or "NO") in the motion, material, and assembly columns. If the part receives a "YES" to any question, enter a "0" in the CFE column. If the answer is "NO" to all three questions, then enter the quantity of parts entered in column 2 in the CFE column.

9. **Value (V):** Some parts have more value than others. For example, an "off-the-shelf" purchased component has more value than a part that must be specially designed for one unique application. Elimination of high value parts should be carefully considered. Assign part value according to the following scale:

0   New designs, separate fasteners, separate operations and adjustments
1   Previously designed and tooled part
2   External standard part ("off-the-shelf" purchased part)
3   Internal standard part (building block part used in several products)

10. **Manufacturability Rating (MR):** Rate the manufacturability of the part as follows:

Rating	Condition
+	Part is designed for manufacture • Part design minimizes tooling cost • Part design minimizes material cost and waste • Part design minimizes processing cost • Net shape process with minimum secondary processing • Process is appropriate to economic scale
0	Part design and manufacturing process are appropriate
-	• Part is excessively difficult or costly to manufacture • Part is incompatible with manufacturing process • Process is inappropriate for economic scale • Material is inappropriate

11. **Undesirable Interaction Rating (UI):** Rate the potential for undesirable interactions as follows:

Rating	Condition
+	Interactions are "decoupled" • Functions are separated • Part is exactly constrained • Critical dimensions are integrated into one part • Deformations are matched • Precision is appropriate to stiffness
0	No potential for undesirable interactions noted
-	Assembly involves an undesirable interaction • Assembly process is difficult, slow, error prone • Part is over-constrained resulting in sensitivity to deformation, alignment, tolerance stack-up, etc. • Load paths and geometry interact to produce stress concentration and/or residual stress • Part features interact in critical ways with other components

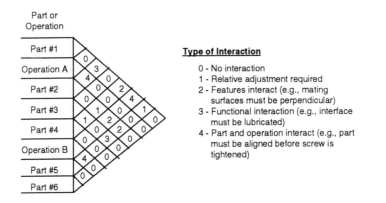

Figure 17.2 Hypothetical example of how an "interaction" matrix could be used to denote interactions between parts in an assembly. Note that, for the interaction notation used in this example, design alternatives having the least information content would have 0s everywhere.

12. **Notes:** Often, there are important aspects of the assembly or a particular part that is not captured by the analysis. The "note" column provides a convenient means for recording this information. Because the column is fairly narrow due to space constraints, it is recommended that the notes be recorded on a separate sheet of paper and referenced on the worksheet by a flag.

17.3 UNDESIRABLE INTERACTIONS

Undesirable interactions can often occur on a variety of levels and may involve several components. These are difficult to capture in Column 11 of the worksheet and it is recommended that a note or other written explanation be used. Alternatively, an "interaction matrix" similar to the house of quality "roof" matrix discussed in Chapter 8 might be used. A hypothetical example illustrating this approach is shown in Fig. 17.2.

17.4 METRICS

The data recorded on the analysis worksheet can be used to generate a variety of different metrics. These metrics measure "before" and "after" designs to gage the amount of improvement in manufacturing and assembly ease. Metrics are also useful for comparing different design alternatives and for benchmarking competitor products. Some of the many metrics that can be computed using the worksheet data include the following:

$$Count\ Ratio = \frac{\sum Qty - \sum CFE}{\sum Qty} \tag{17.1}$$

$$Separate\ Fastener\ Ratio = \frac{\sum Type\ 1}{\sum Qty} \tag{17.2}$$

$$Separate\ Operation\ Ratio = \frac{\sum Type\ 0}{\sum Qty} \tag{17.3}$$

$$Value\ Ratio = \frac{\sum (2\ \&\ 3\ Value\ Ratings)}{\sum Qty} \tag{17.4}$$

Of these, the count ratio, which is the ratio of theoretical parts determined using the three critical questions to total number of parts and operations, is perhaps the most useful general-purpose metric. Note that as information content of the design is reduced, these metrics will approach the following values:

$$Count\ Ratio \rightarrow 1$$
$$Separate\ Fastener\ Ratio \rightarrow 0$$
$$Separate\ Operation\ Ratio \rightarrow 0$$
$$Value\ Ratio \rightarrow 1$$

17.5 SIMPLE EXAMPLE

Suppose we wish to improve the manufacturability of the vacuum cleaner attachment shown in Fig. 17.3. Our starting point is the existing product which we disassemble keeping track of the quantity of parts removed and the order in which they are removed. Numbering the parts in the order they are removed and listing quantity in parenthesis gives the following parts list:

1. Screw (6)
2. Bottom Plate Assembly (1)
3. Brush (1)
4. Spring (2)
5. Label (1)
6. Upper Housing (1)

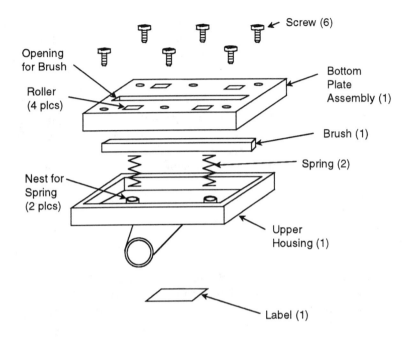

Figure 17.3 Exploded view of a vacuum cleaner attachment for cleaning floors.

Six screws hold the bottom plate assembly to the upper housing. The bottom plate assembly is designated as an assembly because it consists of a plastic injection molded part with several pin-mounted rollers pressed in at different locations to facilitate movement of the attachment over the surface being vacuumed. The spring mounted asymmetrical nylon brush, which is retained between the upper housing and bottom plate assembly, extends through the bottom plate and deflects by compressing the springs to accommodate irregular surfaces. The two coil springs that mount the brush fit into circular fixture nests molded in the upper housing.

17.5.1 Analyze

The manufacturability of the existing design (Fig. 17.3) is analyzed part-by-part as each part is assembled together. This is done by completing a line on the work sheet for each part as it is added to the build and for each separate operation as it is performed. The analysis begins with the *base component* which, for the purposes of the analysis, is defined as the part that is left after the last part has been disassembled. In the case of the vacuum cleaner attachment (Fig. 17.3), the base part is part #6, the upper housing.

To analyze the upper housing, we add the part to the build by imagining that it is loaded "top-down" using one hand into a fixture which holds it so that the next part (i.e., part #5, the label) can be assembled "top-down." We then complete the first line of the worksheet as shown in Fig. 17.4. Note that all of the assembly questions receive a "plus" since the upper housing can be held with one hand, is easy to position and orient, and is added to an ideally designed fixture. Since there are no other parts in the assembly at this stage of the assembly, both the "motion" and "material" questions are answered "no." The "assembly" question is answered "yes" however because a base part is needed for the assembly to be built. Assuming the vacuum attachment (Fig. 17.3) is an existing product, the value of the upper housing is assigned a "1" since it is an existing part. Also, no manufacturability or undesirable interactions are noted.

A similar analysis is performed for each subsequent part as it is added to the build (Fig. 17.4). In doing this, we note the following:

- Because the label is flexible and sticky on one side, it receives a "negative" for each assembly question. Also it is identified as a CFE.
- The reorientation operation is required so that the springs can be added "top-down." Note that the assembly questions do not need to be answered for separate operations and that all separate operations are CFEs.
- The coil springs receive a "negative" to the handling question because they tend to nest and tangle. They receive a "plus" for insertion because they are located in fixture nests molded in the upper housing. A "no" is assigned to the motion question because one end of the spring is stationary. Finally, because they are purchased "off-the-shelf" parts, the springs have relatively high value.
- The brush is difficult to handle because it is asymmetrical and the correct orientation is not obvious. It is also difficult to insert because holding is required.
- The bottom plate assembly is difficult to insert because there is no guiding feature on the injection molding to guide the bristles of the brush through the opening. Also, the "assembly/service" question is answered "yes" because the part is a subassembly and because it must be separate to facilitate assembly of the springs and brush.
- Since the screws are separate fasteners, they need not be analyzed for ease of assembly. Separate fasteners are also automatic CFEs.

The analysis is completed by calculating appropriate metrics. In this example, we calculate the count ratio to be 0.23 (Fig. 17.4).

1	2	3	Assembly				Part Elimination				Assessment			
	2	3	4	5	6	7	8				9	10	11	12
Part or Operation	Qty	Type	H	I	S	C	Motion	Mat'l	Ass'y	CFE	V	M	UI	Notes
Upper Housing	1	2	+	+	+	+	No	No	Yes	0	1	0	0	
Label	1	2	-	-	-	-	No	No	No	1	1	0	0	
Reorient	1	0								1	0			
Spring	2	2	-	+	0	0	No	No	No	2	2	0	0	
Brush	1	2	-	-	0	0	Yes	No	No	0	1	0	0	(a)
Bottom Plate Ass'y	1	3	0	-	0	0	No	No	Yes	0	1	0	0	(b)
Screw	6	1								6	1	0	0	

$$\Sigma Qty = 13 \qquad \Sigma CFE = 10$$

(a) Holding required
(b) Brush hard to guide

$$Count\ Ratio = \frac{13-10}{13} = 0.23$$

Figure 17.4 Manufacturability analysis of the vacuum cleaner attachment shown in Fig. 17.3.

17.5.2 Create

Manufacturability of the design is improved by seeking creative ways to harvest the manufacturability improvement opportunities identified in the analysis. The goal is to redesign to eliminate CFEs and to design the parts that remain to be easy to assemble and manufacture. To fully explore the range of improvement opportunities, it is best to develop several redesign proposals. As a general approach, proposing a "practical" redesign, a "stretch" redesign, and a "radical" redesign is a good way for exploring the design space. A *practical* redesign is one that can be implemented fairly easily without excessive tooling costs or development effort and with no potential change to product performance. A *stretch* redesign, on the other hand, could involve significant development time and cost and may also impact performance to some extent. A *radical* redesign is a stretch redesign usually involving a basic change such as switching to a totally different physical concept or to a totally different material or manufacturing process. To illustrate, suppose we use the manufacturability

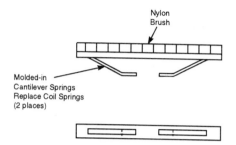

Figure 17.5 The two separate helical coil springs are eliminated by molding cantilevered beams on the brush.

analysis (Fig. 17.4) to propose the following vacuum cleaner attachment redesigns.

Concept A:
- Eliminate the label by integrating it into the upper housing as a "hot stamping" or by making it a part of the plastic injection molded housing.
- Replace the screws with snap fittings. This can be done by making relatively simple changes to the existing upper housing and bottom plate molds.
- Provide a tapered guiding feature on the bottom plate molding to guide the brush bristles through the opening. Again this involves only minor changes to the existing tool.
- Provide a simple snap-fitting in the upper housing to retain the brush thereby avoiding the current instability and holding requirement.
- Assemble as a "Z-axis" sandwich.

Concept B:
- Same as concept A, plus,
- Integrate the two coil springs into the nylon brush by molding in two cantilever springs in the plane of the brush (Fig. 17.5).

Concept C:
- Combine the bottom plate into the upper housing and eliminate the pin-mounted rollers.
- Integrate the brush with a transverse cantilever spring and snap fitting (Fig. 17.6).
- Assemble by snap-fitting the brush to the upper housing.

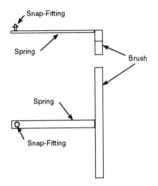

Snap-Fitting

Spring

Brush

Spring

Snap-Fitting

Figure 17.6 Integrated brush and spring with snap-fitting for mounting on the upper housing.

In considering these concepts, we see that concept A can be implemented with relatively little cost and effort. Concept B, on the other hand, would likely require relatively expensive new tooling and could effect product performance. Concept C represents a radical departure from current practice. Performance could be significantly effected because the rollers have been eliminated and the brush may not be as effective because it is less constrained torsionally. Also, integration of the bottom and upper housing will likely require an expensive tool having complex side action.

It is also important to note that, in creating concept C, the team had to re-evaluate the "yes" answer to the "assembly/service" question for the bottom plate assembly (Fig. 17.4). This illustrates a subtle danger associated with the three basic questions. Often a "yes" to one of the questions can be converted into a "no" by considering a different physical concept or working principle. In other words, even when a part has been determined to be a "theoretical" part based on the three basic questions, it should still be challenged. This is especially true if the part is considered to be a theoretical part because of a "yes" to the "assembly/service" question.

Once several alternative redesigns have been proposed, they can be analyzed and compared by completing an analysis worksheet for each proposed redesign. The analysis worksheet for redesign concepts A, B, and C are shown in Figures 17.7 through 17.9. In addition, the existing design and redesign proposals can be compared in a variety of ways. One approach is to compare part counts as shown in Table 17.1. Note that concept C is the low information content solution since it integrates the brush and spring and also eliminates all the parts, material handling, factory floor space, quality risk, and other information content associated with the bottom plate subassembly.

1	2	3	Assembly				Part Elimination				Assessment			
			4	5	6	7	8				9	10	11	12
Part or Operation	Qty	Type	H	I	S	C	Motion	Mat'l	Ass'y	CFE	V	M	UI	Notes
Upper Housing	1	2	+	+	+	+	No	No	Yes	0	1	0	0	
Spring	2	2	-	+	0	0	No	No	No	2	2	0	0	
Brush	1	2	-	+	+	+	Yes	No	No	0	1	0	0	
Bottom Plate Ass'y	1	3	+	+	+	+	No	No	Yes	0	1	0	0	

$$\sum Qty = 5 \qquad \sum CFE = 2$$

Figure 17.7 Analysis worksheet for concept A.

1	2	3	Assembly				Part Elimination				Assessment			
			4	5	6	7	8				9	10	11	12
Part or Operation	Qty	Type	H	I	S	C	Motion	Mat'l	Ass'y	CFE	V	M	UI	Notes
Upper Housing	1	2	+	+	+	+	No	No	Yes	0	1	0	0	
Brush	1	2	+	+	+	+	Yes	No	No	0	1	0	0	
Bottom Plate Ass'y	1	3	+	+	+	+	No	No	Yes	0	1	0	0	

$$\sum CFE = 3 \qquad \sum CFE = 0$$

Figure 17.8 Analysis worksheet for concept B.

1	2	3	Assembly				Part Elimination				Assessment			
			4	5	6	7	8				9	10	11	12
Part or Operation	Qty	Type	H	I	S	C	Motion	Mat'l	Ass'y	CFE	V	M	UI	Notes
Housing	1	2	+	+	+	+	No	No	Yes	0	1	0	0	
Brush	1	2	+	+	+	+	Yes	No	No	0	1	0	0	

$$\sum CFE = 2 \qquad \sum CFE = 0$$

Figure 17.9 Analysis worksheet for concept C.

Table 17.1 Comparison of Alternative Redesign Concepts

Design Alternative	Part Count	Parts Eliminated	Sub-Ass'y Eliminated	Count Ratio
Original Design	12	---	---	0.23
Concept A	5	7	0	0.60
Concept B	4	8	0	1.00
Concept C	2	10	1 (13 parts)	1.00

17.5.3 Winnow, Refine, and Optimize

In the "winnow, refine, and optimize" phase, the performance and cost implications of each alternative must be evaluated with respect to customer driven performance and product cost targets, and with respect to development time and resource constraints. In addition, "hybrid" concepts that combine the best features of each design proposal should be evaluated. For example, it may be possible to replace the integrated spring/brush design used in concept B (Fig. 17.5) with the integrated spring/brush design used in concept C. Also, pin mounted rollers can be added to concept C to provide desired functionality.

17.6 KEY TAKEAWAYS

Manufacturability of a product is improved by reducing the information content of the design.

- The most effective way to reduce information content is to eliminate parts and make those that remain easy to assemble.
- Part count is reduced and product manufacture and assembly simplified by following design for efficient manufacture guidelines.
- The guidelines can be systematically applied using a disciplined analyze, create, and refine methodology.

18

Elimination and Simplification Strategies

18.1 INTRODUCTION

In this chapter, we use the guided common sense approach to explore a variety of strategies for reducing information content of the manufacturing system as a whole. Recall from Chapter 6 that guided common sense is essentially the use of common sense to *eliminate* sources of information content and to *simplify* by reducing the amount of information contained in the information sources that remain. Therefore, the goal of each strategy is to eliminate and/or simplify some aspect of the manufacturing system.

Because the manufacturing system as a whole often involves business considerations as well as many different organizations within the firm, implementation of many of the strategies discussed require company wide buy-in, participation, and management support. Implementation will also likely require significant resources and planning. However, when successfully implemented, the benefits of improved profitability and business performance that they are likely to produce make them well worth the effort and expense.

18.2 USE GROUP TECHNOLOGY

Group technology (GT) is an approach to design and manufacturing that seeks to reduce manufacturing system information content by identifying and exploiting the sameness or similarity of parts based on their geometrical shape and/or similarities in their production process. GT is implemented by utilizing classification schemes to identify and understand part similarities and to establish parameters for action. Manufacturing engineers can decide on more efficient ways to increase system flexibility by streamlining information flow, reducing setup time and floor space requirements, and standardizing procedures for batch-type production. Design engineers can develop an attitude of designing for producibility and help eliminate tooling duplication and redundancy.

18.2.1 Grouping Schemes

The grouping of related parts into part families is the key to group technology implementation. The family of parts concept not only provides the information necessary to design individual parts in an incremental or modular manner, but also provides information for rationalizing process planning and forming machine groups or cells which process the designated part family. A *part family* may be defined as a group of related parts that have some specific sameness and similarities. Design-oriented part families have similar design features, such as geometric shape. Manufacturing-oriented part families share similar processing requirements. In principle, part families can be based on any number of different considerations. For example, parts manufactured by the same plant, or parts that serve similar functions such as shafts or gears, or parts all fabricated from the same material could conceivably be grouped into part families.

18.2.2 Methodology

In implementing GT, the problem that immediately presents itself is, how are the parts to be effectively grouped into meaningful part families? Three methods are commonly used: (1) visual inspection, (2) production flow analysis (PFA), and (3) classification and coding. The first method is obviously very simple, but limited in its effectiveness when dealing with a large number of parts. Production flow analysis is a technique that analyzes the operation sequence and the routing of the part through the machines in the plant. The three steps involved in PFA are factory planning, process planning, and operation planning. Parts with common operations and routes are grouped and identified as a manufacturing part family. Similarly, the machines used to produce the part family can be grouped to form the machine group or cell. For PFA to be successful, it must be assumed that the majority of the parts belong to clearly defined families and the machines to clearly defined groups. One of the advantages of this technique is the ability to form part families without using a classification and coding system. Part families are formed using the data from operation sheets or route cards instead of part drawings. There are also a number of disadvantages associated with the PFA approach. Chief among these are its reliance on existing production data and routing methods and the difficulties involved in manually sorting the production data.

Part classification and coding is perhaps the most effective and widely used method. In this approach, parts are examined abstractly to identify generic features that are captured using an agreed upon classification and coding system. The main advantage of classification and coding is the accuracy it provides in forming part families. The major disadvantage is cost. Classification and coding systems are expensive to develop in-house or to purchase from the outside. In addition, employees must be trained and the actual coding of the existing parts is labor intensive and time consuming. Additional costs are incurred to achieve data retrieval and analysis. If purchased outside, the system may need to be modified to fit the environment. Finally, there is the cost of associated hardware and its operation. Offsetting this direct cost is the enormous cost savings potential through the exploitation of similarity to reduce information content that group technology makes possible.

Although a number of commercial classification codes are available, none can be considered universally applicable. The best coding system is one that is adapted specifically to the industry or company where it is used. Ideally, a coding system for classifying parts should allow the designer to visualize both the part and its process plan by the code number alone. The level of coding sophistication is one of the first decisions that the user of GT must make. Shorter codes, which are easier to use and less prone to error, are also less flexible. For computerized GT systems, the simplest scheme is often a hard-coded program, in which an interactive routine asks the user a series of questions regarding a particular part. More sophisticated applications use soft-coded programs, such as decision-tree coding, where the length of the code depends on which branch of the logic tree is being used.

18.2.3 Classification Codes

There are two coding approaches that are commonly used: attribute-based (also referred to as polycodes) and hierarchical-based coding (also referred to as monocodes). In attribute coding (Fig. 18.1a), the simpler of the two, code symbols are independent of each other. Hence, codes of fixed length span all part families and each position in the code corresponds to the same variable. Because of this, each attribute that is to be coded must be represented by one digit, which can make the code quite long in some cases. One advantage of the attribute-based code is that information can be extracted quickly from the entire parts population. In addition, it is relatively simple to develop. Unfortunately, the price paid for such elegance and simplicity is an extremely long code when many parts families are involved.

A hierarchical code structure is designed so that each digit in the sequence is dependent on the information carried in the digit just preceding it (Fig. 18.1b). Generally, the first digit holds the most basic information, and each succeeding digit contains more specific information. This makes it possible to

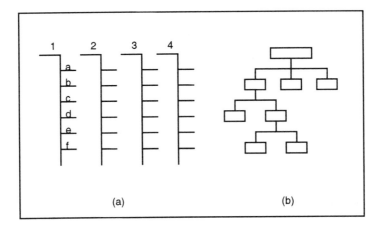

Figure 18.1 Two commonly used methods of coding: (a) attribute-based and (b) hierarchical-based.

capture a large amount of information in a relatively short code. A disadvantage is that the entire code must be interpreted because no one digit carries any significant information. Monocodes are also more difficult to store in databases that do not use a decision-tree approach.

Because both methods have advantages and disadvantages, many modern coding systems rely on a hybrid type of code. One useful compromise is a fixed-length code that is standard for all part families for the first few digits and variable for the rest. A major characteristic of today's GT software programs is the ease with which they can be modified and customized. This is a necessity because each application must be adapted to suit each company's part type.

Codes can be configured to capture a wide range of information. Basic information includes part geometry, dimensions, mechanical characteristics, and manufacturing features. Specific variables may include shape, finish, and size. More sophisticated codes also capture subelements (such as holes, cones, slots), material type, chemistry, finish, and special processes such as heat treatments. Production scheduling can be facilitated by including lot sizes, information on order frequency, relationships between processing steps for particular parts, and assembly or quality control requirements.

In recent years, computer programs using cluster analysis, pattern recognition, and other methods have been developed to enhance the conventional methods of PFA and classification and coding for more effective grouping of part families. Also, while traditional classification and coding systems are still useful and well adapted, new concepts for generative coding and computer-oriented schemes are becoming possible with modern CAD

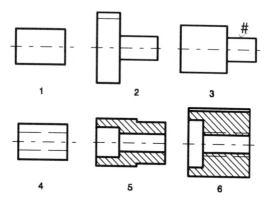

Figure 18.2 Family of rotational parts.

system developments. One of the promises of these developments is the possibility of associating a database with the solid model of a part. If properly implemented, such a database could eventually facilitate the same part grouping capabilities that is currently provided by the part-coding scheme.

The benefits derived from group technology are not limited to design and manufacturing economy. With increasing emphasis on flexible and integrated manufacturing, GT is proving to be an effective first step in structuring and building an integrated database. Standardized process planning, accurate cost estimation, efficient purchasing, and assessing the impact of material costs are benefits which are also often realized. The competitive edge which group technology can provide when teamed with a company's computer, can be significant, and should be considered along with other high-technology productivity enhancement equipment such as robotics, CAD/CAM/CAE, and MRP systems.

18.2.4 Simple Coding Example

Consider the family of rotational parts shown in Fig. 18.2. For this family, all external cylindrical surfaces are finish turned, the stepped surface on part #3 is finish ground, and the counter bore on part #5 must be finish turned (bored). Suppose we wish to develop a coding scheme that captures the part geometry and all manufacturing information required to machine the part family from cut bar stock. One possible coding scheme is shown in Fig. 18.3. Note that this is a "hybrid" code having three fixed digits. The first two digits are assigned according to particular part attributes. The third digit is assigned depending on the answer to a series of questions. Using this scheme, each part in the part family of Fig. 18.2 would be coded as shown in Table 18.1.

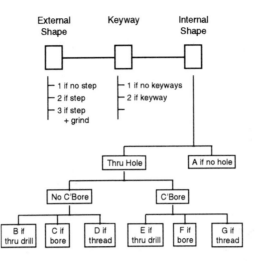

Figure 18.3 One possible classifiction code for capturing the features of the family of rotational parts shown in Fig. 18.2.

18.3 USE "PREVIOUSLY DESIGNED" PARTS

Instead of designing a new part from scratch, time and effort can often be saved by modifying an existing part design. When this is possible, design of a new part becomes a matter of loading an existing solid model or CAD drawing and then changing dimensions and deleting or adding features as required. With the availability of parametric solid modeling software, the process can be as easy as changing values in a parameter table.

Using previously designed parts as a starting point for new designs has been widely used for the design of parts such as chain drive sprockets, V-belt sheaves, shafts, valve bodies, sheet metal enclosures, heat exchangers, and so forth because only dimensions and detail features typically change in new designs. Using this strategy as a general approach is more challenging, however, because the designer must be able to quickly identify and locate the right starting part. In some companies, this isn't a problem because the variety of different parts is small enough that most of the company's design engineers are familiar with all available choices. As the number of existing designed parts increases, however, a point is rapidly reached where it is easier to start from scratch than to look for an appropriate "previously designed" part.

One way to help solve this problem is to use group technology (GT). In this approach, a classification code is used to capture part similarities (see previous section). To use the GT system, the design engineer identifies the code that

Table 18.1 Part codes

Part Number	Code
1	11A
2	22A
3	31A
4	11B
5	21F
6	12G

describes the new part he or she wishes to design. A search of the GT database then quickly reveals whether one or more similar parts already exist. If similar parts are found, and this is most often the case, then the designer simply reviews these parts and selects the one that is closest to the part to be designed.

In large companies, it is not unusual to discover many versions of the same general part. For example, there may be several to several hundred different "shaft" designs used in a large company such as an agricultural equipment manufacturer. To avoid having to review all of these parts to identify the best starting part, it is often convenient to group similar parts into part families. A *master part* for each family can then be created by consolidating all of the features present in each version of the part into one *composite* design (Fig. 18.4). Note that the master part may be a hypothetical part if no existing part has all the features of the part family.

A general methodolgy for using previously designed parts is as follows:

1. Sort all existing parts into logical part families. In large companies, this is often best done using a GT classification code.
2. Create a master part for each part family.
3. Create a CAD library of master parts.
4. To design a new part, copy the most appropriate master part from the master part library and modify as required to create the new part.

Obviously, many variations of this procedure are possible depending on the particular situation involved. The ultimate goal is to make it easy for the designer to begin with an existing design rather than creating the geometry for a new part from scratch each time. How this is done depends on the specifics of each unique situation. It is particularly important to note that a formal GT system is not mandatory for implementing this strategy. In many companies, part families are obvious and can be quickly identified by inspection alone. In others, especially those producing a large variety of different products, a formal GT system may be the only practical approach.

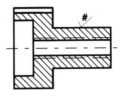

Figure 18.4 Composite part for the part family shown in Fig. 18.2.

In addition to saving design time, using previously designed parts to design new parts produces the following benefits:

- Gradual evolution of "repeat parts."
- Gradual standardization of part families.
- Improved cost estimation.
- Improved ease of manufacture.

Repeat parts are parts that are used for different purposes in different products. For example, a mounting plate or bracket used to mount a particular component in one product can be used to mount the same or a different component in another product. Similarly, a spacer can also serve as an axle, or lever, or standoff elsewhere. Repeat parts are usually "discovered" rather than intentionally designed because it is often difficult to imagine how the same part can be used in different products and in different applications. Using previously designed parts as a starting point for new designs facilitates this discovery process by associating different and often totally unrelated applications with similar parts.

Another benefit is the gradual standardization of part families that is likely to occur. By always using the same set of starting parts, designers will naturally tend to simplify the design task by using the same fillet radiuses and other detail features in all new designs. Over time this practice will result in fewer variations and in more standardized part family features. In some instances, it may be possible to accelerate this tendency, or even better, to optimize the results by implementing a long-range standardization plan. Goals of such an approach should include the following:

- Minimize the number of different part families.
- Minimize the number of variations within each family.
- Minimize the number of design features used in each variation.

Using previously designed parts also facilitates improved cost estimation because actual cost for the previously designed part is known. Finally, ease of manufacture is improved because tooling and production experience gained from the previously designed part can be factored into the new design in a continuous improvement process.

18.4 REDUCE PROCESS COUNT AND PROCESS TYPES

An unnecessarily large number of different manufacturing processes, or number of individual processing steps, or number of different tools and fixtures, or processing complexity increases information content. Design decisions that reduce these counts and that favor less complex, widely used processes will eliminate information content. For example, the choice of unit manufacturing processes usually depends on material selection, production volume, and the final properties and characteristics required for the part. To eliminate sources of information content, the number of different processes and materials considered should be limited to those that are most widely used within the company. If this is not possible, then the part should be purchased from a supplier who is skilled in the particular process of interest and is properly equipped to make high quality parts in the quantity required.

In general, the use of near net-shape processes are preferred whenever possible. Likewise, secondary processing (finish machining, painting, etc.) should be avoided if possible. One way to avoid secondary processing is to specify tolerances and surface finish that are compatible with the capabilities of net-shape processes such as precision casting, P/M, or plastic injection molding. Also, material alternatives that avoid painting, plating, buffing, and other surface treatments should be considered. This recommendation is based on the recognition that higher material cost and/or direct cost can be accepted if it leads to lower total cost.

In addition, only those joining processes and assembly processes that are widely used within the company should be considered. Don't use rivets if spot welding is used to assemble all other products. It is often advisable to avoid processes such as adhesive bonding where several processing steps are required and final properties are dependent on hard-to-control factors such as curing time and quantity of adhesive applied. If such processes are necessary for functionality or performance, then they should be treated as a core expertise within the company and given proper technical and financial support.

18.5 DEVELOP A ROBUST DESIGN

Variability and randomness are enemies of manufacturing because they add information content to the manufacturing system. Extra steps and activities are required to correct problems, rework and repair defective product, process warranty and customer complaints, and so forth. The result is poor quality, unnecessary manufacturing cost, product unreliability, and ultimately, customer dissatisfaction and loss of sales. Variability reduction and robustness against variation of hard-to-control factors are therefore recognized as being of paramount importance in the quest for increased profitability.

Design decisions, made in the early stages of design, can often reduce and possibly eliminate the information content associated with variation and randomness by making the product inherently robust. A *robust design* is insensitive to and/or tolerant of hard-to-control variation and is therefore, by definition, a low information content design. In developing a robust design, the design team seeks to maximize functionality while simultaneously minimizing the effects of hard-to-control variation. This has the effect of maximizing the "signal to noise ratio" of the design, where the signal is a measure of how effectively a product satisfies its intended function and noise is a measure of disturbance factors which detract from or degrade functionality.

Several strategies for developing a robust design are discussed in the following. It is likely that all of these strategies can be employed to varying degrees depending on the product or system being developed

18.5.1 Avoid Undesirable Interactions

Detrimental effects of hard-to-control variation can often be avoided by "decoupling" interactions between various aspects of a design. The seal design illustrated in Fig. 18.5 is a case in point. Because relative motion occurs between the seal and the rotor surface, most of the frictional heat generated due to rubbing will flow into the rotor causing it to expand. In Design A, this hard-to-control thermal expansion will result in a thermo-instability since contact force and resultant frictional heating will increase with rotor expansion. In Design B, the undesirable interaction is avoided since rotor expansion will decrease contact force and frictional heating to produce thermo-stable behavior.

18.5.2 Specify Robust Parameter Values

If one were to plot the relationship between particular outputs of a product or system and the design variables that effect them, nonlinear relationships would often be discovered. In particular, regions are likely to be found where hard-to-control variation produces little or no variation in the desired output or performance of the system (Fig. 18.6). This insight leads to a second way of optimizing robustness, which is to specify numerical values for design variables that minimize sensitivity to hard-to-control variation. Robust parameter values can be found by performing design sensitivity studies using design of experiment (DOE) methods such as the Taguchi method. Probabilistic design

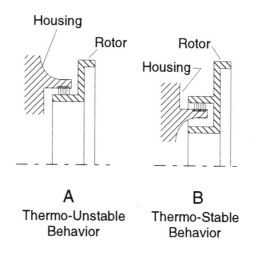

A
Thermo-Unstable
Behavior

B
Thermo-Stable
Behavior

Figure 18.5 Avoiding undesirable interaction in a rotating seal design (Pahl and Beitz, 1988).

can also be used when the design relationships are modeled mathematically. These methods are discussed further in Chapter 21.

18.5.3 Eliminate Sources of Variation

Variation can occur in many forms in the product function and on the manufacturing floor. Eliminating sources or causes of variation and error is therefore an effective means for implementing robust design (Fig. 18.7). Flexible components are a major source of variation. For example, assembly of dangling connectors, lead wires, and other unrestrained parts is difficult to automate because dealing with location and orientation randomness requires human intelligence. In addition, quality risk for these types of parts is increased because their location can vary from product to product and is not controllable. These difficulties are avoided by carefully planning the layout of electrical connections early in the design and by providing features such as recesses, channels, wire guides, and clamps that locate, orient, and restrain the wires and connectors. Complex wire runs, potentially dangerous "pinch points," and location uncertainties are avoided. The amount of wire required is reduced. Otherwise arbitrary design decisions such as the location of electrical connectors are constrained. Delicate wires are protected from chafing and accidental damage. Potential wiring errors are avoided. Information content of the design is reduced. *Variation is eliminated by designing to make manufacture and assembly easier to "do right" than to "do wrong."*

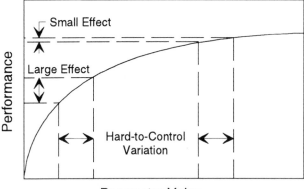

Parameter Value

Figure 18.6 Parameter design exploits nonlinearity between product functional characteristics and hard-to-control variation.

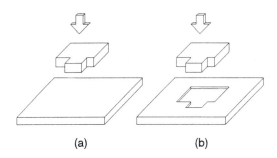

Figure 18.7 Design parts to be self-locating and self-aligning. (a) This part could be placed in any location and orientation. (b) This part has a nest to help locate and orient.

18.5.4 Design-In Variation Tolerance

Another approach is to provide features that are variation tolerant. For example, providing a feature that is easily recognized by a vision system, regardless of lighting conditions, makes the vision system less sensitive to variations in lighting conditions. Similarly, use of generous tapers and other guiding features makes part insertion less sensitive to placement accuracy of an assembly robot. Use of selective compliance can be an effective way to relax tolerance requirements (Fig. 18.8). Many design for assembly considerations also help to make product assembly variation tolerant (see Chapter 13).

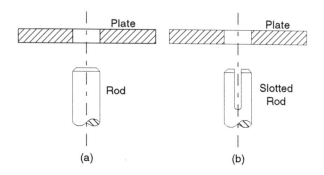

Figure 18.8 Providing compliance in selected directions can offset the need for tight dimensional tolerances. (a) Obtaining a tight fit between the solid rod and plate requires tight tolerances and close control of interference. (b) The slotted rod can deform slightly in the lateral direction permitting looser tolerances and more interference for a tighter fit (Chow, 1978).

18.6 MAKE PRODUCTS AND PARTS SMALL

Small products and parts require less material and therefore material cost, as well as material handling and shipping cost, is less. They also weigh less, and occupy less space making them less costly to fabricate, assemble, handle, and pack because smaller, lighter duty equipment and tooling can be used. Small volume, low weight, and a small "footprint" are also highly desirable from a product use and customer satisfaction standpoint. For these and many other reasons, miniaturization is an extremely potent simplification strategy. Miniaturization, however, almost always requires a total redesign of the product. It also must be technologically feasible and cost effective to implement. Experience has shown that part elimination, especially by using integral design and by identifying designs that require fewer theoretical parts (see Section 13.4), often leads to smaller product size. Note, however, that assembly information content can increase if parts are too small (see Section 13.5.1 for guidelines).

18.7 MAXIMIZE COMMONALITY

Information content is greatly reduced when separate parts or aspects of the product possess common features or attributes. For example, if several fasteners must be used, each fastener should be the same, or if this is not possible, the number of different sizes and variations should be minimized. Similarly, designed components and subassemblies should have common interface features and attributes (e.g., location, configuration, mounting details, and detail geometry). The same is true for interconnections (e.g., electrical, hydraulic, pneumatic) between components and subassemblies.

Common interface geometry and design details simplify by making frequently occurring features of the product everywhere the same. This reduces the information content of all aspects of the product's design and manufacture. Solid models (or drawings) are more quickly created by cutting and pasting or by using libraries of common geometry details. Tooling, assembly processes, inspection processes, and so forth is simplified by replacing variety with repetition. Servicing is simplified, again, because repetition and sameness is substituted for variety. Supplied component procurement is simplified because interface details are clearly specified and widely understood.

To maximize commonality, planning must begin in the very early stages of design. Often, by planning the interconnection, interfacing, and mounting details of the "black boxes" (e.g., designed components, subassemblies, and modules) before they are designed, detail design of each black box is facilitated because the location of connectors and other interface details are no longer arbitrary decisions. Commonality should be used at all levels of the product. Use common purchased components such as washers and ball bearing, use common geometric features and dimensions on all designed parts, use common manufacturing processes and tooling to produce the parts.

18.8 COORDINATE THE PRODUCT AND PROCESS

Simplify the manufacturing system by designing the product and process as a coordinated system. For example, designing guiding and self-locating features into the product components and using a simple SCARA (selective compliance assembly robot arm) type robot for component insertion involves less information content than using a multi-degree-of-freedom robot combined with tactile sensing and vision. In doing this, strive to distribute information content wisely across all elements of the manufacturing system including the designed components, tooling, and automation hardware, software, and "human ware." If an operation or particular aspect of the product is particularly difficult to automate, for example, the low-information content solution may be to perform the operation manually rather than trying to design this aspect of the product to be automation friendly.

Ultimately, simplification of the manufacturing system is maximized when complete coordination between the product and the manufacturing process is achieved. Doing this requires careful and extensive planning and organization supported by clearly articulated business, marketing, and manufacturing strategies. It also requires a concurrent engineering approach. See Chapter 12 for a general approach for coordinating the product and process.

18.9 DESIGN FOR SERVICEABILITY

A product that is easy to manufacture and assemble, will in general, also be easy to service and maintain. However, additional information content can be removed by paying special attention to the servicing and maintenance requirements of the design. Easy service and maintenance encompasses several considerations: (1) reduction of the number and variety of tools, operations, and labor skills required for service and maintenance, (2) ease of accessibility, and (3) ease of reparability.

For ease of servicing and maintenance, it is especially important that design alternatives be selected that avoid the need for special or unusual labor skills, special or time consuming operations, and special tools and fixtures. Ideally, only a minimum number of readily available tools should be required, and all operations should be easy to perform without special training or skills. *The best way to achieve this goal is to clearly define permissible tools and operations early in the design, before detail design decisions are made.*

Ease of accessibility implies easily identified parts that are readily visible and easily reached. In particular, critical parts, components, and structural sections, which are likely to have above average failure rates, should be easy to see and inspect. All maintenance items and points should also be easily accessible. Appropriate self-diagnostics which trip visible warning indicators when malfunctions or failures occur should be used for especially critical components, especially if it is impossible to make the critical component accessible. In designing for accessibility, close attention should be paid to providing sufficient hand and tool manipulation clearance for scheduled maintenance items, adjustments, and so forth, without removal of interfering components. Sufficient clearance should also be provided for removal of components, modules, and subassemblies without removal of additional components. Use of solid modeling capabilities, which permit simulation of assembly and disassembly operations, can be particularly useful for insuring that sufficient clearance is provided.

Reparability is optimized by maximizing commonality and interchangeability of components, providing readily available and easily used troubleshooting aids, minimizing and simplifying adjustments, and designing to minimize repair complexity and turnaround time. Problem isolation can be greatly simplified by utilizing individual module diagnostics and providing visual inspection means such as sight glasses, indicators, pointers, gage ports, manual overrides, and so forth. To facilitate module exchange or repair, modules should be designed with self-guidance features for quick and easy removal and installation. Components should also utilize quick automatic disconnect and self-guidance features. Whenever possible, common mounts, connections, threads, connectors, fasteners, belts, tubes, hoses, filters, lubricants, etc. should be used. Information content will be a minimum when a short list of acceptable common components is agreed upon prior to detail design and then used exclusively in the design. Such an approach not only simplifies and improves reparability, but also tends to maximize availability of spare parts and enables parts "scavenging" and substitution in emergencies.

18.10 DESIGN FOR RECYCLABILITY

The ability to recycle easily and efficiently is becoming an increasingly important consideration in product design. The information content involved in recycling can be reduced in several ways: (1) recycling by repair, (2) recycling by conversion, (3) design for easy separation, and (4) contamination tolerant design (Chow, 1978).

In the recycling by repair approach, modules and products are designed so that they are easy and cost effective to repair and/or rebuild. This makes it economically feasible to repeatedly refurbish and reuse them as opposed to disposing of them. The use of rebuilt alternators and starter motors is common practice now. Eventually, factory rebuilt and reconditioned items may be used in new products. For example, refurbished radiators, seats, engines, transmissions, and other automotive components might be used in new automobiles. To make this economically attractive, such components must be designed in a way that makes them easy to disassemble and rebuild. As a minimum, it must be possible to restore all wear points to original specifications without costly processing or the need to replace expensive parts.

Recycling by conversion involves reusing items in different ways after their intended function has been fulfilled. Using a jelly container as a drinking glass is an example. Another example is Henry Ford's use of the wood from shipping containers as floorboards in the Model T. A somewhat related example is the use of scrap or byproduct from one manufacturing process as the raw material for another. Sawdust and wood chips are used in this way to make particleboard and other building products.

Designing products so that material separation is easy is another approach. One method is to make easily removed components such as automotive radiators out of a single material, or to use an integrated wiring harness so that copper can be quickly reclaimed. Alternatively, markers or tracers embedded in the wanted materials can be used to activate sensors in the sorting process. Finally, parts can be designed to be tolerant of contamination. A black plastic part can accommodate more reclaimed material because color degrading can be easily concealed. The same is true for painted or metalized plastic parts.

18.11 DESIGN FOR THE ENVIRONMENT

Product design is the first and most important step in creating an environmentally friendly product life cycle. Product design decisions determine how the product is manufactured, distributed, used, and ultimately disposed. Most importantly, design decisions made about functionality, product configuration, and material specifications constrain downstream processes that often limit the freedom the firm has for reducing environmental impact of its operations. *Design for the environment* is a design strategy that seeks to reduce information content by anticipating Health, Safety, and Environmental (HS&E) impacts of a new product design. Doing this requires a coordinated understanding of marketing, design, packaging, and manufacturing issues. It also requires an understanding of government regulations and compliance requirements. The following design for the environment checklist is useful for exploration of HS&E issues (Allied Signal, 1994).

- How can current products and processes be improved with respect to environmental issues? Are the environmental issues affecting the customer clearly understood?
- If higher prices result from environmentally conscious design and manufacturing, have the reasons for cost differentials been communicated to distributors and customers?
- Have customer preferences for environmentally safe products and processes been identified? Have focus groups, surveys, and customer advisory panels been used to identify customer environmental requirements?
- Have recycling and reuse systems that will increase customer willingness to recycle or reuse products been identified?
- Can recycled materials be substituted for primary raw materials?
- Which materials used in the product or process are hazardous or regulated? Can they be eliminated?
- Has the cost of safety equipment, training, and exposure monitoring been considered in the materials selection process?
- Have all legal and regulatory requirements been considered?
- Will use of the product by the customer require excessive energy consumption or generate air, water, or soil pollution?
- Has packaging been eliminated where possible? Has the remaining packaging been designed to be reusable? If not, is it environmentally friendly? Is it biodegradable, compostable, or recyclable?
- Does the packaging provide instructions for proper use? Are instructions for reuse, return, or recycling clearly marked? Have systems for the return and reuse of packaging material been created?

- Have production processes been designed to minimize use and release of toxic or hazardous materials?
- Have production lines been evaluated and optimized for efficient energy usage?
- Has the cost of waste and scrap disposal been considered?
- Have reclamation, recycling, and reuse systems been included in the process design?
- Does the process design maintain segregation of waste streams?

18.12 DESIGN FOR MATERIAL HANDLING

Material handling involves the flow of materials to, from, and within the manufacturing workspace. Material handling adds no value and can be a major source of information content. Design decisions often determine what material flows are necessary for product manufacture and how they must be implemented in the plant. By considering material flow early in the design process, some of these flows can be greatly simplified or possibly eliminated altogether. Material flows to be considered include the following:

1. *Product Flow*: Assembly process path along which the unfinished product travels.
2. *Workspace Flow*: The handling that occurs within the workspace directly prior to assembly.
3. *Supply Flow*: The direct movement from a receiving process to marshaling area or a workspace with a one-to-one matching between product flow requirements and supply flow requirements.
4. *Hardware Flow*: The direct movement of components from a marshaling area to a workspace. Similar to supply flow except pieces are not individually accountable because of small size and value.
5. *Trash or Scrap Flow*: The movement of scrap (especially used packaging materials) away from the workspace to a marshaling area or the plant boundary. Rejected and/or defective materials are included in this flow.
6. *Bulk Material Flow*: The movement of loose, unpackaged materials without constant dimensions to the workspace.
7. *Container Flow*: The movement of reusable shipping materials, both containers and dunnage, from the workspace back to a marshaling area or to the plant boundary.
8. *Fixture Flow*: The movement of assembly fixtures, both for subassemblies and for the product, back to the initial workspace for reuse.

In seeking to eliminate and simplify material handling, the design team should carefully consider how the product design influences each of these material flows. The following material handling checklists are useful for guiding an exploration of material handling issues. Note that a workspace flow checklist is not provided since this flow is discussed extensively in Chapter 14.

18.12.1 Questions Common to All Flows

- How many moves are required? How can these be reduced and/or simplified?
- Are any operations (e.g., inspection, orientation, sequencing, etc.) performed during transit? How can these be reduced and/or simplified?
- How will the material flow system cope with changing product complexity and marketing changes?
- How will the material flow system cope with model mix changes?
- How will the material flow system cope with technology changes, either in assembly strategy or material flow equipment?
- How can the product, component, or part be better designed to more easily interface with and/or simplify this flow?

18.12.2 Product Flow Checklist

- How flexible must the product flow system be? When will different product types be carried on the system: concurrently during a single shift, on a second shift, or after completion of a model run? How rapid must changeover be?
- What transport fixture will be required to provide access to the installation points yet securely hold the product? Will more than one type of fixture be required? If so, how will an incomplete product be transferred from one fixture to the next?
- What happens to a unit with a problem (e.g., parts that don't fit, or a missing part)? Will the defective unit be taken out of the flow system? If so, what will be the mechanism of removal, temporary storage, and return to the flow line?
- What product design changes would simplify the product flow system?

18.12.3 Supply Flow Checklist

For each supply flow:
- Is there an orientation that must be maintained?
- Are all components adequately secured?
- To what extent must components be protected from damage?

- Is the component unusually heavy, bulky, or valuable?
- If the component chosen for a specific production unit must be chosen by production lot or serial number, is it easy to identify? How does the operator know that the right component has been installed?
- What product or component design changes would simplify or eliminate this supply flow system?

18.12.4 Hardware Flow Checklist

For each hardware flow:
- Must the hardware be loaded into a bin/hopper/bowl at the workspace? If so, will the components be able to withstand associated handling forces and abuse? Are they free of defects that could interfere with part feeding?
- Is the hardware prepackaged with an orientation that must be maintained?
- Are special tools required? How and when are these supplied to the operators?
- What product or component design changes would simplify or eliminate this hardware flow?

18.12.5 Trash or Scrap Flow Checklist

For each scrap or trash flow:
- Must some scrap materials be segregated from others, either due to their hazardous nature or the fact they will be reworked, recycled or sold?
- Are special containers required for some types of scrap?
- Is the scrap stored in reusable containers? If so, their movement is similar to container flow.
- Can scrap materials be compacted (e.g., collapsed, nested, etc.) to reduce the volume transported?
- Is defective material returned to the supplier? If so, how much must be saved to make a shipment? Where will it be held?
- What product or component design changes would simplify or eliminate this scrap flow?

18.12.6 Bulk Material Flow Checklist

For each bulk material:
- Are special tools and applicators required? How and when are these supplied to the operators?

- Does the bulk material have a shelf life or setup time that must be monitored?
- What product or component design changes would simplify or eliminate this bulk material flow?

18.12.7 Container Flow Checklist

For each reusable container:
- Is it really cost-effective to return the chosen container to the supplier? Would a different container be more efficient? What container features would enhance its performance or usefulness?
- Do similar containers returning to the same supplier location stack, nest without jamming, interlock, or collapse for more compact transport?
- Are inserts (if any) captured for loss prevention?
- Can container sizes and shapes be standardized throughout the plant and ultimately throughout the company?
- What product or component design changes would simplify container function, facilitate standardization, or eliminate the need for this container?

18.12.8 Fixture Flow Checklist

For each assembly fixture:
- How many different types of fixtures are being used in the same fixture flow? How can they be differentiated, stored, and reissued? How can they be combined, simplified, standardized?
- Do fixtures require special power supply considerations (e.g., connect, disconnect, temporary plugging, etc.) while being transported?
- Are unique rework fixtures required to move rejected stock or subassemblies?
- What product or component design changes would simplify the fixture function, facilitate standardization, or eliminate the need for this fixture?

18.13 HARMONIZE THE PRODUCT MIX

Many mature companies supply a full line of products to satisfy a range of market needs. Over time, this can lead to a proliferation of product models and variants that have been designed to meet a variety of narrowly defined needs. Some products among this array typically sell well while others hardly ever sell, resulting in a mix of high and low volume products. When this condition exists, we refer to the low sales volume products as "oddballs" and "specials."

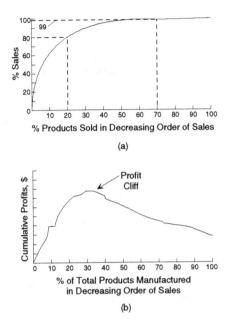

Figure 18.9 Total cost of oddballs and specials often exceeds the sales revenue they generate. (a) A sales-volume profile such as this suggests that traditional costing methods may be giving a wrong cost picture. (b) Oddballs and specials can reduce profitability because information content is largely independent of sales volume.

Oddballs are regular products (i.e., are stocked, listed in sales literature, advertised, etc.) that have relatively low sales volume compared to other, more popular, product models. *Specials* are modifications and adaptations of regular products that are created to meet unique customer needs.

To illustrate the effect of oddballs and specials on profitability, consider a manufacturing business that produces a large number of different products. Suppose that the sales-volume profile for this firm is as shown in Fig. 18.9a. In this hypothetical case, approximately 80% of the company's sales are generated by 20% of its products. In addition, about 70% of the company's products account for more than 99% of sales. Hence, the remaining 30% of products are generating less that 1% of the sales volume.

How is this mix effecting profitability of the business? The answer depends on how overhead cost is related to sales volume. If overhead cost is proportional to direct cost, as is implied by the way many firms assign overhead charges, then low volume products can appear to be profitable since direct cost is a function of volume. However, based on our understanding of information content, we know that much of the information content associated with a product is independent of volume. If this is the case, then "oddballs" and "specials" can be much more costly than traditional cost estimation practice would indicate.

If we assume that information content is independent of sales volume, then profitability of a particular product can be calculated by the expression

$$Profit \: / \: Product = V \: (Selling \: Price) - [V \: (m + l + p) + AI] \qquad (18.1)$$

where V is the sales volume, m, l, and p the material, labor, and production cost per unit, respectively, I the total information content of the product, and A the \$/bit proportionality constant that relates information content in bytes to cost (see Section 6.3). Using Eq. (18.1), cumulative profit for the sales-volume profile given in Fig. 18.9a is plotted as shown in Fig. 18.9b. With information content assumed to be independent of sales volume, the oddballs and specials are seen to drastically reduce profitability by creating the "profit cliff" shown in the figure. This result has been confirmed by using ABC to analyze costs in companies having a mix of high and low volume products (Kaplan, 1990).

The profit cliff associated with a mix of high and low volume products can be avoided in three ways:

1. Determine what the "oddballs" and "specials" actually cost (direct + indirect) to produce and charge accordingly.
2. Eliminate "oddballs" and "specials" from the product lineup.
3. Design the product and/or the manufacturing process to make overhead cost more proportional to direct cost.

Although all three alternatives are viable business options, the third option is of special interest since it is addressed by many of the topics discussed in this book. In particular, we see that sales volume is an important consideration in determining how a product line should be designed to maximize profitability. For products enjoying high sales volume, direct costs are likely to dominate so these products should be designed to minimize material and labor cost. Also, in general, high volume products benefit from more complex, integral part designs, especially if they reduce overall part count and material volume.

For oddballs and specials, indirect cost dominates, so these products should be designed to minimize information content associated with their diversity and with the effort required to customize and manufacture them. Often, designs that allow "mixing and matching" of standardized modules and building blocks result in the least information content because they combine flexibility with increased production volume for individual components. See Chapter 20 for further discussion of modular and building block designs.

When there is a high degree of diversity and a mix of high and low volume products, it may be impossible to achieve one highly profitable design. In these cases, the low information solution may involve two separate product designs, one for high volume products and the other for low volume products. Alternatively, different manufacturing methods may be considered. For example, suppose a foundry produces a broad range of cast products and that the large number of oddballs and specials produced is creating a "profit cliff" such as that shown in Fig. 18.9b. Faced with this situation, one alternative the foundry might consider is to continue using the casting process to produce the high volume products and to replace casting with CNC machining to produce the low volume oddballs and specials.

As a closing remark, it should be noted that the concept of "oddballs" and "specials" exists at all levels of manufacture. A seldom-used drill bit size, a fastener used in just one low volume application, an oddball connector that has to be special ordered are all examples. Eliminating these sources of information content over time will produce significant total cost reductions.

18.14 KEY TAKEAWAYS

Total cost of a product is determined, in large measure, by the amount of information content associated with the product's design, manufacture, operation, and life-cycle support. The less information content, the less total cost. Therefore, design strategies that focus on reducing the information content of the manufacturing system as a whole have great potential for minimizing cost. In developing and implementing elimination and simplification strategies, the design team should:

- View the manufacturing system, including all activities associated with new product introduction, day-to-day production, sales, installation, operation, maintenance, service and life-cycle support, as an enormous collection of information sources.
- Realize that each product design decision, both large and small, contributes to the number of information sources and to the amount of information contained within each source.
- Strive to make design decisions that limit the number of different types of information sources and the number of individual sources contributing to each type while also avoiding undesirable interactions.
- Endeavor to make design decisions that reduce or minimize the amount of information content of each individual source.

19

Standardization Design

19.1 INTRODUCTION

The goal of *standardization design* is to systematically identify and evaluate standardization opportunities within the firm that have high cost savings potential. Standardization is the most far-reaching and all encompassing approach available for reducing information content of a product design. Standardization works by limiting the number of options and simplifying the options that remain. It can be applied to components, products, materials, manufacturing processes, business processes, and organizational procedures.

In addition to the many benefits that standardization offers, standardization may also require compromise that can restrict design and marketing options in undesirable ways. It can also be very costly to implement. Therefore, to identify and select promising opportunities, the tradeoffs between benefits and cost should be understood and evaluated. In this chapter, we explore the benefits and costs of standardization and present a systematic process for navigating the major issues of standardization.

19.2 METHODOLOGY

Is standardization the right course of action and if so, what should be standardized and how should the standardization be structured? This is the central question that must be answered when contemplating the viability of a standardization opportunity. Although standardization reduces information content, it does not make sense for every application. To be viable, the benefits must usually far outweigh the costs. In addition, even when the cost/benefit ratio is very favorable, the opportunity is still not viable unless the firm can afford the investment required. The challenge, therefore, is to identify and structure standardization opportunities in a way that maximizes the benefits while simultaneously minimizing the cost.

These insights provide the motivation for our approach to standardization design. It consists of the following major activities.

1. **Understand.** Understand all aspects of the business to identify promising standardization opportunities.
2. **Evaluate.** Develop a qualitative understanding of the costs and benefits associated with each candidate opportunity.
3. **Refine.** Develop a quantitative understanding of the costs and benefits associated with each candidate opportunity as necessary. Use this understanding to winnow, refine, and structure the candidate opportunities into a short list of the most promising and prioritize based on total cost savings potential.

Understanding how standardization could benefit the firm and identifying the opportunities that exist is the essential first step. It helps management to decide if standardization is a wise course of action. It also provides the insight needed for activities 2 and 3 of the process. To surface potential opportunities, the team identifies logical groupings of parts or products that may be standardized. Each logical grouping is referred to as a *candidate opportunity*. It is important to note that the candidate opportunities could involve raw materials, in-process parts, or finished goods. The second activity involves an evaluation process in which the details of each candidate opportunity are fleshed out so that benefits and costs can be listed and qualitatively assessed. In the third and final activity, the most promising candidate opportunities are further analyzed by developing quantitative estimates of benefits and costs. These results provide the insights needed by management to decide the best course of action.

19.3 IDENTIFY CANDIDATE OPPORTUNITIES

Standardization opportunities exist wherever there is excessive information content. Extra information content exists in the form of option complexity, manufacturing complexity, management complexity, inventory complexity, supply chain complexity, and so forth. Therefore, to surface opportunities for standardization, the team should begin by analyzing complexity issues in the business or manufacturing system. Business and manufacturing conditions that signal excessive information content that could benefit from standardization include:

- Extensive proliferation of product models, designed parts, and purchased components.
- Lengthy testing and evaluation cycles for new product designs.
- Excessive inventory, long manufacturing lead times, inability to meet demand, scheduling difficulties, and other operational problems.
- Large numbers of different components purchased in small lot sizes.
- High overhead relative to competitors.

Once sources of extra information content have been identified, the team seeks ways to simplify by using standardization. This involves identifying logical groupings of raw materials, tools, parts, part features, components, or products. Typically, a *logical grouping* will involve some sort of similarity or commonality. For example, a logical grouping may be products that have a similar specification or functionality. Or it could be parts that are manufactured in a similar way or that fulfil similar needs or that have similar geometrical features or shapes. The goal is to discover logical groupings that offer realistic standardization opportunities.

In some situations or industries, logical groupings of products may be obvious or readily identified. In other more complex manufacturing environments, it may be necessary to search the company's database of parts and products to find logical groupings. Unless the database is properly organized, however, such searches can be difficult or impossible. In these cases, the database may need to be reorganized or group technology techniques discussed in Chapter 17 must be used.

As in many other aspects of product design, it is essential in standardization design that a large number of candidate opportunities be identified. By exploring the full spectrum of opportunities across the company, it is possible to begin to see how standardization in one area might positively or negatively effect standardization in other areas. Also, by identifying a large number of different candidate opportunities, the probability of surfacing those opportunities that offer the greatest cost savings potential or quickest payback is greatly increased. Remember also that the candidate opportunities are only ideas at this point, so it is important to defer judgement regarding there cost savings potential or feasibility until the next step in the process.

19.4 EVALUATE QUALITATIVELY

After a range of candidate opportunities have been identified, it is necessary to sort through them and evaluate their total cost savings potential. To do this, the potential benefits and costs associated with each candidate opportunity is listed and the overall potential for total cost reduction assessed on a qualitative basis. Often, this evaluation can be assisted by using the concept of guided common sense discussed in Chapter 6 as well as many of the other methods and practices discussed throughout this book. Also, it is important at this stage to look for synergism as well as undesirable interactions that may occur if more than one of the candidate opportunities is implemented. If the qualitative study shows that a candidate opportunity is worth pursuing, it is retained for further consideration. Those opportunities that are not carried forward are either discarded or retained for future consideration. If many worthwhile opportunities are identified, it may be useful, at this point, to attempt to winnow the less promising and to prioritize those that remain based on potential benefits and costs.

19.4.1 Identify Benefits

The total cost savings potential of a candidate opportunity is assessed by listing its inherent benefits and costs. Table 19.1 illustrates the usefulness of listing benefits and costs. By examining the benefits and costs for a particular practice, it is possible to gain immediate insight into the tradeoffs that exist. Also, the process of identifying benefits and cost teaches much about the standardization opportunity.

Table 19.1 shows that the benefits of standardization result from reduced information content. For example, setups are an extremely information intensive activities that are reduced or eliminated by standardization. Therefore, to identify potential benefits, the team should evaluate each candidate opportunity to determine how the proposed standardization will reduce information content. Each way in which information content is reduced defines a benefit.

In looking at Table 19.1, we see that some of the major sources of information content that are reduced by standardization include the following:

- Setups
- Inventory
- Material Resource Planning
- Design and Development Cost
- Product Options
- Ripple Effects
- Supply Chain

To help guide the benefit identification process, we briefly review how standardization reduces information content of these information sources.

Table 19.1 Benefits and costs of some standardization practices

Practice	Benefit	Cost
Limit the number of product models (e.g., the Ford Model T)	• No setups • Less design time and cost • Economies of scale • More management focus	• Product may be overdesigned and costly • No incentive for repeat buyers to upgrade
Limit the number of product options	• Fewer setups • Less design time and cost • Less equipment cost • Reduced inventory • More management focus	• Product may be overdesigned • May lose customers if options are not available
Limit the number of different feature sizes (e.g., limit hole sizes)	• Fewer setups • Fewer tools	• May have to redesign some products • May have to outsource non-standard sizes
Postpone product differentiation (e.g., customize at regional distribution center to meet regional market needs rather than building many different models at one location)	• Inventory reduction • Decreased shipping cost and lower tariffs by adding value locally • Better customer service • Mores sales	• Basic product may be more expensive due to added features which allow field customization • Distribution centers need personnel, expertise, and equipment
Limit the types of processes and equipment used (e.g., only use spot welding and buy all spot welding equipment from one vendor)	• Reduce the range of training, and specialized skills required • Reduce spare parts inventory • Reap benefits of being a preferred customer	• Lost capability may add cost and/or compromise functionality • Can't meet some needs

Table 19.1 Benefits and costs of some standardization practices (Continued)

Practice	Benefit	Cost
Use existing high volume parts in new, low volume products.	• Save design time • Save tooling cost	• Part may be over designed • May require extra parts to adapt to new product
Rationalize purchased components and use in new product designs (e.g., reduce the number of different standard screws used)	• Less design time • Tested and proven parts and processes • Established suppliers, favorable prices • Interchangeability between products	• Some components may be better and therefore more expensive than needed
Standardize manufacturing processes (e.g., standardize the die shut height of large stamping presses)	• Less setup time • Faster setup may lead to reduced inventory and improved JIT performance	• Cost required to change existing tools and equipment to conform with the standard

Setups

A *setup* includes all of the non-value-added activity and material cost associated with changing a manufacturing process or operation from one product to another. Most setups involve adjustments, tool changes, material purges, scrap, and product rework, all of which generates direct labor and equipment cost. Setups also generate indirect cost due to lost production, lost equipment utilization, and scheduling effort. Standardization avoids setups by allowing equipment to be dedicated to one product.

Inventory

The inventory held by a firm can be divided into two categories: working stock and safety stock. *Working stock* is the amount of inventory held because of the lot size. *Safety stock* is the amount of inventory held to hedge against uncertainty. Standardization may significantly reduce both of these inventories. To illustrate potential reductions, Watson (1996) models the case where one standardized product is used to replace several nonstandard products. Results show that reductions in inventory of up to 50% are possible.

Material Resource Planning

Suppose a firm sells a line of similar but different products. To ensure its ability to deliver each product in a timely manner, the firm must stock the specific parts used to build each individual product. This requires that the firm forecast the demand for each product before the needed parts are ordered. Suppose now that the firm standardizes the parts so that any product in the line can be manufactured using the standardized parts. With standardization, the firm now has the flexibility of deciding which end product the parts will be used in after they are ordered. This added flexibility produces two beneficial effects: improved forecasting and shorter lead times. Forecasting is improved because the firm needs to only forecast demand for the whole product line, not individual products. Lead time is reduced because the larger volume generated by standardization allows more efficient production methods and, since the same parts are used in a variety of products, there is increased probability that parts will be available when an order for a specific product is received.

Design and Development Cost

A product line composed of similar but not identical products can greatly increase fixed costs associated with the design, production, and life cycle support of each individual product. Each product may require its own set of tooling, its own design engineers, its own specialized maintenance and service procedures, its own testing and regulatory approval, and its own suppliers. Standardization will eliminate much of these cost duplications. It will also reduce design and development cost. When a firm uses standard parts, the designers become knowledgeable about the strengths, weakness, performance, and characteristics of those parts. This knowledge facilitates faster and more effective design cycles.

Product Options

To be competitive, many firms must manufacture a full line of products to meet all customer needs. This often leads to extensive manufacturing complexity, especially when the demand for some product models differs markedly from others (see Section 18.13). For example, the General Electric operation that produces panel board components for electrical distribution equipment supplies 40,000 different customer options (Chapter 7; Ettlie and Stoll, 1990). Often, when there is a large number of product options, management might be tempted to reduce this complexity by reducing the number of options. This can be a subtle trap because reducing the number of options does not necessarily mean cost will be reduced, especially if the same parts are required to produce the options that remain. It is therefore extremely important that real cost savings be tied to any proposed reduction in product options.

As discussed elsewhere in this book, the real key to reducing the cost associated with product options is to reduce the number of parts. This is illustrated by the GE panel board example cited above. By developing a standardized "building block" design, the number of different parts required to produce the 40,000 different product options was reduced from 28,000 unique parts to 1,275 building block parts. This resulted in a double-digit percentage reduction in overall cost. In addition, the cycle time for shipping a customer order was reduced from two weeks to three days, two of which are spent getting the order from the customer to the appropriate starting point in production. Product models are reportedly changed a dozen or more times each day and the plant has improved productivity by 250 percent. Backlogs in the plant have gone from two months to two days.

Ripple Effects

The benefits of standardization ripple in many ways throughout the manufacturing system. For example, many suppliers offer a quantity discount since larger orders reduce the supplier's costs by reducing setups and allowing the use of more efficient, less costly production methods. If the standardization involves a supplied part or component therefore, a quantity discount may be an added benefit. However, it is important that the actual cost savings be quantified before counting this as a benefit.

Another ripple effect is cost reduction due to economies of scale. By reducing the number of options, volume for a few parts will be greatly increased. If these parts are produced in-house, this will, in turn, allow the use of automation and more efficient manufacturing processes that will generate significant cost savings. If the part is purchased from a supplier, these savings will be passed on to the supplier. In this case, it should not be automatically assumed that this is a benefit unless a profit sharing arrangement with the supplier has been negotiated.

Supply Chain

Standardization benefits often cascade through the supply chain. That is, the benefits of standardization that are passed on to the firm's suppliers are, in turn, passed on to their suppliers. The whole supplier network should therefore be considered when analyzing potential benefits of standardization.

19.4.2 Identify Costs

Standardization is not free. Table 19.1 shows that profits may suffer because something extra is "given away" (e.g., unutilized material or functionality) or because the product doesn't satisfy certain customer wants or needs. Excessive capital investment may also be required to implement the standardization.

Overdesign

A standardized product must still meet the needs that were being met when several different products were used. As a result, it may be necessary to "overdesign" the standardized product. Such products are *overdesigned* in that they are capable of better performance or more functionality than a particular application requires. A major concern is the tendency for overdesigned parts to weigh and cost more because they usually are designed for the "worst case" situation. Also, as production volume increases, so does the cost penalty for using overdesigned parts. Hence, overdesign generally is less acceptable for high volume products. It is important to note, however, that in some situations, it is possible for the cost of an overdesigned part to be assigned more importance that it may actually merit because the extra material cost and/or manufacturing cost is usually easy to estimate.

There are also harder to quantify costs that must also be considered. For example, the extra weight may result in less efficient product operation or reduced performance. Such penalties can produce a serious or not so serious competitive disadvantage depending on the importance of these characteristics to the customer. In other products, especially those where the overdesigned parts facilitate a desirable improvement in performance and functionality, the "upscale" products may increase in popularity. In these cases, the cost penalty imposed by the overspecified parts could eventually be offset as customer demand for the higher margin upscale product increases.

Based on this discussion, there appears to be at least two clear cut opportunities for reducing the cost of overdesign:

1. Standardize in ways that don't result in overdesigned parts.
2. Standardize in ways that increase sales demand for the least overdesigned products.

Customer Needs

Products proliferate to meet customer needs. If standardization causes the firm to exit some market niches or causes the customer to prefer a competitor product, revenue may be lost. This can be a severe cost penalty if the product is not standardized wisely. Therefore, the firm should carefully analyze market niches and customer desires to correctly understand how standardization will impact sales revenue. In some cases, the standardization will go totally unnoticed by the customer. For example, it is unlikely that the end customer will notice or care that the same screw is used everywhere in the product. On the other hand, if all product models look alike, incentive to purchase the "top of the line" product model may be reduced.

Capital Investment

Depending on the particular standardization scheme involved, capital investments may need to be made. For example, capital investment may be needed to design the new standard product. New tools and equipment may be needed. New plants and warehouses may be needed. Employee training or new worker skills may be required. For the candidate opportunity to be feasible, the savings produced by standardization must recoup all of these costs. If the capital investment is very high, the payback period may extend over many years.

19.5 EVALUATE QUANTITATIVELY

Up to now, most of the benefit and cost analysis has been qualitative. In some cases, this may be enough to demonstrate the viability of a particular candidate opportunity and further analysis is not needed. For most cases, however, additional quantitative data will be required to answer questions, alleviate uncertainties, and generate sufficient confidence for management to make a decision. Generating these numbers is the objective of this activity.

Typically, insights gained from the qualitative evaluation will help guide the detail quantitative analysis performed in this step by providing an awareness of the key variables and relationships that need to be studied. These insights will also guide the team in performing sensitivity studies and making the necessary assumptions to perform the analysis. Obviously, the actual analysis that is performed will depend on the particular standardization opportunity being investigated. For some cases, a simple spreadsheet model will be sufficient, while for others, a more elaborate simulation model may need to be developed.

Performing detail quantitative analysis teaches much about the opportunity. Like "just build it" models (see Chapter 11), detail analysis answers questions, raises new questions, and provides insights that can not be gained in any other way. With these new questions and insights, it is also frequently necessary to revisit other earlier steps of the process to clarify and develop additional information. When each step is executed in a thorough and conscientious manner, however, the process will usually converge fairly quickly, even when some iteration is necessary.

19.5.1 Analysis Questions

Watson (1996) suggests that the following questions be used to guide the quantitative assessment of standardization benefits.

- How much production time is lost while setting up? How much is that time worth? What real dollar savings can be realized by eliminating the setup and gaining production time?
- How much extra labor is needed for the setup? How much is it worth? What real dollar savings can be realized by eliminating the setup and avoiding the labor?
- Are there scrap and other costs generated each time a setup is performed? What real dollar savings can be realized by eliminating the setup and avoiding the scrap generation?
- Are there other ways to reduce setup costs besides standardization? This question serves as a reminder that standardization may not always be the best answer to a problem.
- How much inventory could be realistically eliminated with standardization?
- What is the true cost of holding inventory (value of capital, damage while in storage, value of floor space, etc.)?
- What strategic role does inventory play in other policy decisions? How will standardization effect this role?
- Which raw materials, parts, or components have the longest lead time? Is standardization an option to help improve these lead times?
- Have the cost of lost sales, the cost of markdowns, and the accuracy of forecasts been measured?
- Are demands easy to forecast, difficult to forecast, or both? How would standardization improve forecasts and reduce cost of lost sales and markdowns?
- Will tooling be eliminated and simplified when products are standardized?
- Will there be less overall design work with standardization?
- Will life-cycle support be easier with standardization?
- Will product testing and debugging be easier with more standardization? What is the value of these activities? How soon will the real dollar savings be realized?
- Will standardization lead to real dollar savings or just a reduction in numerical complexity?
- Will standardization reduce the need for management overhead? Will this result in real dollar savings?
- Will standardization help the firm focus on its strategic needs? If so, how will this be measured?
- Will standardization help improve throughput, equipment utilization, customer satisfaction, and productivity? Will it reduce lead time?
- By buying from fewer sources, will there be better overall prices?

- Will standardization generate a quantity discount? If so, how will the unit price for each item change?
- Are there hidden costs or risks associated with depending on only a small number of suppliers?
- Would a single supplier have lower unit cost by producing all the products?
- Are suppliers willing to share the savings generated from increased economies of scale?
- Can the supplier savings be clearly demonstrated and put into a contract?
- Is the supplier relationship good enough to reap some of the non-contractible benefits of supplier reduction?
- Is the firm in a position to examine supplier operations in search of common benefits?
- When the parts are standardized, does the efficiency of the supplier's operations improve? If so, is the supplier willing to share the savings with the firm?
- Do any of the previously mentioned savings opportunities apply to the entire supply chain?
- At what point in the supply chain are products customized? Could this point be delayed? If so, what would be the savings? And, what would be the cost of delaying customization?

Questions to ask when analyzing standardization costs include:

- Will the standard product cost more to produce than each of the individual products it replaces?
- Will the burden of meeting many new specifications add cost?
- Will the product mix change in the future to reduce or increase the costs associated with overdesigned parts?
- Will the customers care about or notice the standardized product?
- How many customers will switch to another product or competitor product because of the standardization?
- If the standardized product is an upgrade for essentially the same price, how many customers, if any, will be lost?
- Will the firm gain new customers by offering a less expensive or less complicated product line?
- Will the manufacturing process change when products are standardized? Will this create required new investments?
- Is it more difficult to design a standard product?

19.5.2 Winnow, Refine, Promote

The quantitative analysis should provide a clear picture of the cost/benefit trades as well as a good "feel" for the promise of each candidate opportunity considered. With this insight, the team is in a position to formulate and prioritize various standardization projects and to make recommendations to management. In promoting the selected candidate opportunities to upper management, remember that identifying promising standardization projects is only the first step in a successful implementation. Standardization generally requires a company-wide vision combined with top-level management support. It also requires careful planning and execution to ensure long-term success. Ultimately, it is often these strategic issues that make or break a good standardization idea. It is therefore extremely important that the team not loose sight of these aspects of standardization design as it seeks to leverage standardization as a product design strategy.

19.6 KEY TAKEAWAYS

- Standardization is the most far-reaching and all-encompassing approach available for reducing information content of a manufacturing system. It works by limiting the number of options and simplifying the options that remain.
- Standardization can be applied to materials, components, products, manufacturing processes, business processes, and organizational procedures.
- The benefits of standardization are numerous and extend to all aspects and levels of the business including the end customer, product design, manufacturing and support operations, and the supply chain.
- The benefits of standardization usually come at a cost. Standardization may require compromise that can restrict design and marketing options in undesirable ways. It can also be very costly to implement.
- To identify and implement successful standardization projects, all benefits and costs should be clearly understood both in qualitative and quantitative terms.

20

Standardization and Rationalization

20.1 INTRODUCTION

Standardization and rationalization (S&R) is an approach that seeks to reduce information content of the manufacturing system as a whole by limiting the number of design choices to a few "best" options. In the S&R approach, *standardization* is the reduction in the number of options (e.g., parts, processes, and so forth) used in *existing* products. *Rationalization* is the identification of the fewest number of "best" options to be used in future products.

Success of the S&R approach depends on only using the rationalized options in new product designs. By doing this, all of the information content associated with numerous options is slowly, but surely, eliminated over time. In many cases, it may take five or more years for the benefits of S&R to accrue, since considerable time is required for old products to be phased out and replaced by new product designs that utilize the rationalized options.

Because of the far-reaching consequences and the significant time and effort involved, S&R requires extensive planning and patience. All organizations involved in new product development including design engineering, manufacturing, service, and purchasing should be involved in the S&R effort from the beginning and should work together to develop a vision for S&R and a plan for implementing it. Without broad commitment and long-term management support, combined with a well thought out rationalization strategy, S&R efforts are almost always doomed to failure.

One of the most fertile areas for the S&R approach is in limiting and controlling the proliferation of purchased components. As reported by Bradyhouse (1987), Black and Decker has used S&R to reduce the number of plain washers purchased by the company worldwide from 448 to 7 (one material, one finish, one thickness). Similarly, the original list of 266 ball bearings has been rationalized to 12 (one seal, one lubricant, one clearance, metric only). In addition to the favorable prices and long-term supplier commitments that result from purchasing larger quantities of fewer components, Black and Decker has also experienced some unexpected benefits. For example, design time is shorter because the designer spends less time selecting from long lists of options. Also, less testing is required since the same life parts (ball bearing, switches, etc.) are used widely in other products. Hence, there is a plethora of field data available for use in judging life expectancy, warrantee cost, and so forth.

20.2 METHODOLOGY

Standardization and rationalization can be applied to almost any aspect of a business or manufacturing enterprise. To be successful, a systematic methodology that includes the following activities is recommended.

1. **Organize.** Decide on the scope of the effort. Form a team composed of representatives from all concerned organizations within the company. Obtain appropriate management buy-in and resource commitment.
2. **Research.** Systematically characterize all options currently available. Seek to understand (1) usage patterns, (2) the importance of differences between closely related options, (3) why popular options are popular, and (4) why unpopular options are occasionally used. Often, group technology (GT) and other techniques can be used to assist in this research.
3. **Rationalize.** Develop a rationalized set of options for use in future products that are acceptable to all new product development stakeholders. Analyze the rationalized list of options for all undesirable interactions and consequences. Look several product generations into the future to gage the effect on performance, functionality, and customer satisfaction. Anticipate manufacturing and technology change.
4. **Implement.** Use the rationalized list of options in all new product designs. For long-term success, each organization within the firm, especially those who are associated with new product introduction, must accept and commit without exception to using the rationalized list. This means that each organization understands and accepts that they may have to find creative new design, marketing, and manufacturing solutions or, in some cases, possibly sacrifice a marginal advantage or performance benefit.
5. **Standardize.** Determine the most popular options. Where possible, make design changes to existing products to eliminate unpopular options and expand usage of the most popular options. These popular options may or may not be part of the rationalized options to be used in new product designs.

Table 20.1 Purchased parts list

Category	Brief Description	Number of Parts
Hardware Items	Electrical Connectors	81
	Springs	145
	Cold Headed Fasteners	359
	Rivets	100
	O-Rings	29
	Insulators	234
	Locks and Hinges	22
	Threaded Nuts	62
	Washers	61
	Miscellaneous Items	149
Purchased Parts	Castings	13
	Stampings	104
	Powder Metal Parts	6
	Molded Parts	73
	Packaging	706
	Nameplates	602
	Labels	386
Raw Stock	Multi-slide	25
	Copper (bar, plate, etc.)	232
	Aluminum (bar, sheet, etc.)	63
	Mill Steel (master coils)	40
	Steel (coils, bar, plate, etc.)	72
	Coatings	10
	Molding Powders	43
TOTAL PARTS		3617

20.3 CASE STUDY

To illustrate the application of S&R to purchased components, suppose that one of the long-term goals of a large manufacturing company is to lower its purchasing costs. With this in mind, the director of purchasing forms a team made up of representatives from product engineering, marketing, manufacturing, quality, service, and purchasing.

20.3.1 Organize

As a first step, the team reviews the extensive list of items that are purchased by the company (Table 20.1). In contemplating the variety and number of purchased components, it is obvious that there is tremendous opportunity to reduce both direct and indirect cost by rationalizing the number of purchased parts. At the same time, the task of doing so appears to be overwhelming, because any reduction in options will require "buy in" from product engineering and other organizations such as marketing and application

Table 20.2 Number of purchased parts and direct dollars expended

Category	Number of Purchased Parts	% Total	% Total Direct Dollars Expended
Cold-Headed Fasteners	359	9.9	3.2
Hardware Items	1242	34.3	18.4
All Items	3617	100.0	100.0

Table 20.3 Distribution of usage among different cold-headed fasteners

Number of Different Fasteners	Usage (Fasteners/Year)	% of Total Usage
16 Highest Volume Fasteners	34,557,700	55%
100 Highest Volume Fasteners	30,319,000	96%
359 (All Fasteners)	62,832,000	100%

engineering, who are directly concerned with product function, performance, quality, serviceability and customer perception. It is therefore, also clear that the S&R effort must proceed in a well-planned and organized manner.

Realizing the challenge of S&R, the team decides to limit its efforts to rationalizing cold-headed fasteners. This decision is based on analysis of the various opportunities available, which showed cold-headed fasteners to be an excellent starting point in the overall effort to rationalize purchased parts (see Tables 20.2 and 20.3).

20.3.2 Research

Having narrowed the scope of the effort to cold-headed fasteners, the team next performs research to develop the basis for a rationalization strategy. Goals of the research include the following:

1. Analyze usage patterns of the current options to understand which options are popular and which are unpopular.
2. Understand the importance of differences in closely related options.
3. Understand why popular options are popular.
4. Understand why unpopular options are occasionally used.

In performing this research, it is usually necessary to characterize the variety of different purchased components in terms of clearly identifiable features or attributes and then relate these attributes to usage patterns within the company. How this is done will depend largely on the nature of the

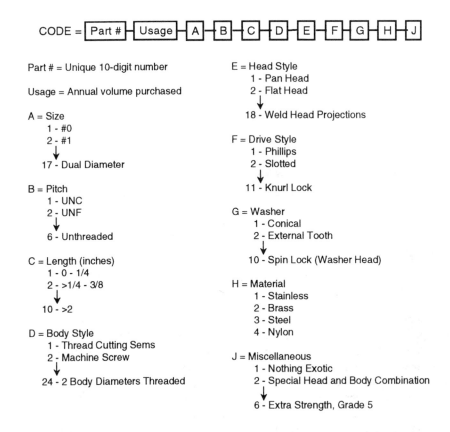

CODE = [Part #]─[Usage]─[A]─[B]─[C]─[D]─[E]─[F]─[G]─[H]─[J]

Part # = Unique 10-digit number

Usage = Annual volume purchased

A = Size
  1 - #0
  2 - #1
    ↓
  17 - Dual Diameter

B = Pitch
  1 - UNC
  2 - UNF
    ↓
  6 - Unthreaded

C = Length (inches)
  1 - 0 - 1/4
  2 - >1/4 - 3/8
    ↓
  10 - >2

D = Body Style
  1 - Thread Cutting Sems
  2 - Machine Screw
    ↓
  24 - 2 Body Diameters Threaded

E = Head Style
  1 - Pan Head
  2 - Flat Head
    ↓
  18 - Weld Head Projections

F = Drive Style
  1 - Phillips
  2 - Slotted
    ↓
  11 - Knurl Lock

G = Washer
  1 - Conical
  2 - External Tooth
    ↓
  10 - Spin Lock (Washer Head)

H = Material
  1 - Stainless
  2 - Brass
  3 - Steel
  4 - Nylon

J = Miscellaneous
  1 - Nothing Exotic
  2 - Special Head and Body Combination
    ↓
  6 - Extra Strength, Grade 5

Figure 20.1 The 359 purchased cold-headed fasteners are characterized using a simple polycode. Note that a particular code value is associated with each possible attribute parameter value. For example, if the fastener is made of brass, the material attribute (H) would be assigned a code value of "2".

components under study. In the case of the cold-headed fasteners, the team is able to employ group technology methods to develop a categorization scheme. *Group technology* (GT) is a technique that seeks to exploit the sameness or similarity or parts (see Section 17.2).

    GT is implemented in this case by first analyzing a sample of the cold-headed fasteners to define the important attributes of the different cold-headed fasteners and then developing a simple code designed to capture relevant information regarding each attribute. The code is summarized in Fig. 20.1. By including the part number in the code, individual fasteners can be traced to specific product applications. This is important for understanding why

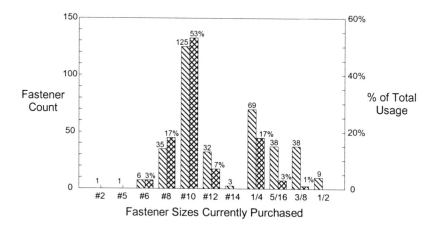

Figure 20.2 Fastener size distribution. Total fastener count = 359; total fastener usage = 62,832,000 per year.

unpopular options are sometimes used. Similarly, the usage field allows the popularity of particular fastener attributes to be determined by associating them with annual volume. A particular fastener is coded by entering its 10-digit part number, the total fasteners purchased per year (usage), and the code value corresponding to the appropriate parameter value for each attribute (A–J). Each of the 359 purchased cold-headed fasteners was coded and the results entered into a computerized database. The database was then used to develop parameter value frequency distributions for each attribute. A typical result is shown in Fig. 20.2.

Results, such as those shown in Fig. 20.2, are then analyzed by the team to understand usage patterns and to develop an understanding of popular and unpopular options. For example, in Fig. 20.2, it is seen that popular fastener thread sizes include #8, #10, #12 and ¼-inch. In addition, the largest size used is ½-inch. This would suggest that the eleven different fastener sizes currently purchased could possibly be reduced to three sizes, #10, ¼-, and ½-inch. Similar analysis is performed on all attribute categories captured by the code.

20.3.3 Rationalize

Using the research results, the team next develops a tentative rationalization scheme, which it then circulates throughout the company for comment. Based on company-wide response to the rationalization proposal, modifications are made and re-evaluated until all stakeholders are satisfied and willing to accept the new, rationalized set of options.

It is extremely important that the team expends the time and effort required to gain complete, company-wide consensus on the rationalization plan. This may require a number of meetings and/or discussions with key personnel in each stakeholder organization. It also requires that the team be very open to suggestions and objections, that they listen carefully, and that they genuinely attempt to reach acceptable compromises with respect to all issues. Most importantly, all stakeholders must be actively involved in the process, otherwise important input will be missed and implementation will be resisted.

In the case of the cold-headed fasteners, the team was eventually able to reduce the 359 different purchased fasteners to a total of 44 (Table 20.4) using an iterative negotiation-compromise-consensus process. This represents an 88% reduction in purchased cold-headed fasteners. Note that the rationalized fastener set is essentially composed of eleven different fasteners, each purchased in a variety of lengths. When fully implemented over time, large savings in direct and indirect cost can be expected due to the reduction in information content produced by the rationalization. Fewer tools are needed for driving fasteners, material handling, storage and retrieval complexity is reduced, quality risk is reduced, and high volume discounts for every fastener purchased become possible.

20.3.4 Implement

The team works with all new product design and development organizations to insure that only the rationalized list of 44 fasteners is used in new product designs. As part of the implementation activity, the team works with management to insure that use of the rationalization scheme is continuously expected and nurtured. The team also educates all employees about the advantages of using the rationalized list, explains how the rationalized list was developed, and discusses the design compromises that may be required. Finally, to ensure that the long-term benefits of the S&R effort are harvested, management must resist granting exceptions, no matter how extenuating the circumstances. Each time an exception is granted, unnecessary information content is added to the manufacturing system and the benefits of the S&R effort are diminished. If exceptions must be granted, this is an indication that the rationalized list is not optimal and a re-evaluation and possible modification to the list should be considered.

20.3.5 Standardize

As a general rule, the rationalized options are to be used only in new designs and no attempt should be made to retrofit existing products with the rationalized options. However, for existing products, it is often possible to replace unpopular options with more popular options by making minor low-cost

Table 20.4 Rationalized cold-headed fastener options

Size	Pitch	Head Style	Drive Style	Length (inches)	Body Style
#8	Plastic	Pan Head	Torx-Slot Combination (T-20)	3/8, 1/2, 3/4	Trilobe Thd. Forming
#10	UNF	Pan Head	Torx-Slot Combination (T-25)	3/8, 1/2, 3/4, 1, 1 ¼, 1 ½	Tri-Point Screw
#10	UNF	Pan Head	Torx-Slot Combination (T-25)	3/8, 1/2, 3/4, 1, 1 ¼, 1 ½	Tri-Point Conical Sems
#10	Plastic	Pan Head	Torx-Slot Combination (T-25)	3/8, 1/2, 3/4, 1	Type 25 Thread Cutter
#10	Plastic	Pan Head	Torx-Slot Combination (T-25)	1	Trilobe Thd Forming
1/4	UNC	Hex Washer Head	Torx-Hex Combination (T-30)	1/2, 3/4, 1, 1 ¼, 1 ½	Tri-Point Screw
1/4	UNC	Hex Washer Head	Torx-Hex Combination (T-30)	1/2, 3/4, 1, 1 ¼, 1 ½	Tri-Point Conical Sems
3/8	UNC	Hex Head	Hex Socket	1, 1 ½, 2, 2 ½, 3	Hex Head Cap Screw
3/8	UNC	Oval Head	Square Shoulder Lock	4 ½, 5 ½, 6 ½	Std. Shoulder Carriage Bolt
1/2	UNC	Hex Head	Hex Socket	1 ½, 2 ½, 3 ½	Hex Head Cap Screw
1/2	UNC	Oval Head	Square Shoulder Lock	1 ½, 2 ½, 3 ½	Std. Shoulder Carriage Bolt

design changes. For example, referring to Fig. 20.2, it may be possible to replace the #2 and #5 size fasteners with a #6 or even possibly a #8 fastener by making simple design changes to the existing products. Similarly, it may be possible to eliminate all #14 size fasteners by replacing them with a #12 or ¼-inch size fastener. Note from Table 20.4 that #6, #8, and #12 fasteners are not included in the rationalized fastener set. However, they can still be used in existing product designs to help eliminate unpopular options and increase the concentration of popular options. This helps realize near-term benefits of S&R while avoiding major redesign expenses.

20.4 KEY TAKEAWAYS

- Standardization and rationalization is a powerful and extremely effective way for reducing the indirect cost of doing business in a manufacturing enterprise.
- S&R can be applied to all aspects of the firm's operation including parts, processes, tools, fixtures, material handling methods, and so forth.
- Perhaps the best place to begin implementing S&R is with off-the-shelf purchased parts, especially hardware items and high volume components.
- Group technology techniques and other methods that help characterize and categorize variety are extremely helpful in understanding most S&R opportunities.
- To be successful, S&R should be performed using a structured methodology and a team approach that includes all stakeholders associated with new product design and development.
- To be successful, S&R must have a strong base of company-wide support. Management must provide vision, long-term commitment, and sufficient resources.
- Benefits of S&R include:

  - Less of everything in the day-to-day operation of the business.
  - Less design time, faster design cycles.
  - Tested and proven parts and processes.
  - Established suppliers and favorable prices.
  - Interchangeability and portability.

21

Internal Standard Components

21.1 INTRODUCTION

In this chapter, we present a variety of strategies for standardizing internal components. The objective is to leverage the benefits of standardization while minimizing its cost by converting designed components into internal standard components. Recall from Chapter 12 that a *designed* component is a unique part or subassembly that is designed in its entirety during the new product design. A *standard* component, on the other hand, is a previously designed part or subassembly that has been designed in such a way that it can be used interchangeably in a variety of different new product applications. Standard, supplied or "off-the-shelf" components such as electric motors, light bulbs, electrical connectors, and mechanical fasteners, are referred to as *external* standard components. *Internal* standard components are standardized parts and subassemblies that are unique to a particular firm, but are used interchangeably in a variety of products and/or product families manufactured by that firm.

Internal standard parts tend to work best in companies that produce and sell a variety of similar or related products and that are constantly introducing new and improved product versions. Hand-held power tools are a good example. Functionally, an electric drill is totally different from an electric sander. From a part decomposition standpoint, however, both products are similar in that they both utilize an electric motor, gears, shafts, sleeve bearings, control switches, and so forth, enclosed in a housing. Many of these parts, such as the shafts and bearing, if standardized, can be used interchangeably in different products. Such standardization produces increased flexibility and economies of scale. Also, time to market is decreased because previous design field experience with these parts reduces the need for extensive design analysis and reliability testing of new products that utilize these parts.

310

In developing internal standard components, it is important to recognize that the parts in most products can be divided into two general groups: unique parts and common parts. *Unique* parts are parts such as housings, brackets, chassis, and labels, which are configured differently and provide different functionality for each particular product family or product model. *Common* parts, on the other hand, are parts such as shafts, bushings, gears, seals, and springs that have similar configurations and serve similar functions in all product models. In an internal combustion engine, the engine block, cylinder head, intake manifold, crankshaft, camshaft, and oil pan are unique parts. Conversely, the pistons, connecting rods, cylinder liners, wrist pins, tappets, valves, and valve springs are common parts.

21.2 DEVELOP A MODULAR PRODUCT

A *module* is a self-contained component that is equipped with standard interfaces that allow it to be integrated into a larger system. Modules form "building blocks" that can be used interchangeably in different products and product variants. For example, different combinations of building blocks can be used to specialize a product to meet particular customer specifications (Product A in Fig. 21.1). Different camera body and lens combinations are an example. Alternatively, different modules can be used interchangeably in different combinations to create different final products (Products A, B, and C in Fig. 21.1). Power tools that utilize different combinations of motor and drive modules are illustrative of this approach. Modules can also be repeated within a product (Product C in Fig. 21.1) allowing a larger system to be built from repetitions of smaller components. Note that this not only increases production volume for the smaller components, it also effectively implements the "simplify by making parts small" strategy discussed in Chapter 18. This approach has been widely applied to modular furniture and cabinets.

Modules can be defined in a variety of different ways. In a "Lego block" building set, each individual part is a module. Many manufactured products such as electric motors, light bulbs, batteries, circuit breakers, heat exchangers, and so forth are also modules when viewed from the perspective of a larger system. At the other end of the spectrum, a product can be organized into major physical groupings, sometimes called "chunks" or physical "building blocks." Each chunk is composed of a collection of components that implement the function of the product. Using the concept of "chunks" in the design of a product can provide many of the advantages of modular design without any evidence of modularity being apparent in the final product. For example, in some front wheel drive automobile designs, the power train (engine, transmission, and front suspension components), rear axle assembly, and car body are each treated as separate chunks (see Fig. 12.5).

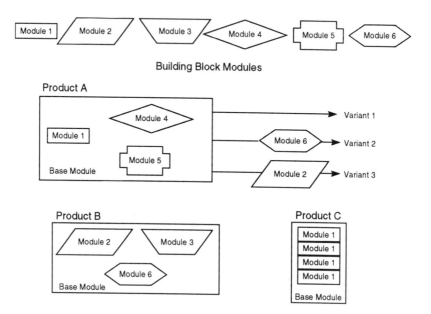

Figure 21.1 Examples of modular design.

A module is sometimes more costly than an integral design because of extra cost of fittings and interconnections. However, when properly implemented, decreased tooling cost and increased production volume associated with modular construction will almost always offset these additional costs. Since a module can be detached and isolated from the rest of the system, malfunction can be easily located, and a failed component can be replaced easily. A module can also be fully checked and tested before it is inserted into, or combined with, the rest of the product. Most importantly, modular design can guard against obsolescence and facilitate a continued program of product improvement since a new generation product can utilize most of the old modules requiring only a few modules to be redesigned or improved. Also, such an approach will significantly shorten the design cycle and reduce the total cost of introducing a new product.

In developing a modular design, a good strategy is to keep the product generic for as long as possible during assembly by saving specialized modules for last. If possible, the modules should be designed to add up to the final product thereby eliminating the need for a housing or other integrating structure. Also, information content is reduced if all modules (except perhaps the base) are approximately the same size.

21.3 CONSIDER STANDARDIZING LARGE PARTS

Large parts can present considerable problems in the design of some products. Lead times are usually long for these parts and the detail design of many other smaller components and subsystems may depend on the detail design of the large part. Similarly, the design of a transmission housing or machine frame often needs to be specified early so that the dimensions and configuration of the overall system can be determined. For these and many other reasons, it is often necessary to firm up the design of large parts early in the program. These early decisions, however, can quickly constrain the design and may possibly add cost to subsequent part designs. Also, should the design of a large part need to be changed, cost and time penalties can be high because tooling and many other parts are likely to be affected.

One way to avoid problems such as these is to consider standardizing large parts. This approach is most feasible in platform products where several different models, usually involving different applications of a proven technology, are involved. How large parts are standardized obviously depends greatly on the particular type of product involved. Often a large part is a base part to which many other parts are added during assembly. Base parts can often be designed as building block parts (modules) by standardizing certain dimensions and features. Also, these parts can sometimes be designed as fixtures, thereby eliminating the need for an assembly fixture and eliminating the material handling required to move the fixture back to the start of the line.

Consider, for example, the design of a laser printer. By using the same size base part for all models, ranging from the slow and inexpensive version for home use to the costly, high speed office version, one standard design can be developed. To maximize the number of variations possible, extra holes, "knockouts," and so forth are included to meet all foreseeable needs and product variations. An added advantage of standardizing the base part is a simplified material handling system. Because the base part is standard and common to all models, the material handling system is independent of (i.e., decoupled from) the particular product or model being built.

In companies having a high product mix, good candidates for standardization can often be identified using group technology (GT) techniques (see Section 18.2). In this approach, large parts manufactured by the company are clustered into part families. Each part family is then evaluated and the composite part for part families which meet required functional and economic criteria are then developed into standard designs (see Fig. 18.4). In one possible scenario, the standardized composite part would be used in all future designs. In another scenario, the manufacturing process might be designed to easily produce desired variants of the part family simply by leaving out various processing steps. Obviously many variations are possible.

21.4 STANDARDIZE UNIQUE PARTS

For most products, unique parts and common parts need to be treated differently. Unique parts tend to specialize or differentiate the product in one way or another. For example, the shape of an electric drill housing facilitates its use. Similarly, the dial of a temperature gage is different from that of an oil pressure gage. Because of inherent differences, it is unlikely that unique parts themselves can be standardized. What can be done, however, is to group unique parts into part families, and use the intrinsic similarities of each part family as a basis for standardization. The goal is to standardize part features and manufacturing processes of unique parts in a way that makes them easy to design, tool, and introduce into production.

To illustrate how a unique part can be standardized, consider the design of internal combustion engine blocks for high volume automotive applications. In general, the engine block must be different for each engine design. The manufacturing process, which consists of first casting the block from cast iron or light metal alloy and then machining detail features, however, is essentially the same regardless of the detail design or end application. In spite of this, it is usually the effort involved in designing, constructing, and proving out the transfer line used to machine the engine block that historically accounts for much of the time and cost required to introduce a new engine.

Can standardization be used to simplify the machining process? Figure 21.2 shows a standardized flexible-machining center for machining automotive engine blocks. Each station consists of a fully programmable multi-axis CNC machining center. The line can be quickly and easily changed over to a new engine block design by simply downloading new NC programs for each station. For this machining solution to work, the engine block must be designed according to the following design rules:

- The envelope dimensions of the block must be within prescribed maximums.
- All features must be obtainable using three orthogonal machining axes.
- To reduce cycle time, holes and other features should be spaced to allow simultaneous machining operations.

Using these rules to guide the design of the engine block essentially standardizes the unique part design. The engine block for each engine application is unique, but each is standard in that it conforms to the above rules. By conforming to these rules, the block can be quickly and easily introduced into production simply by reprogramming the flexible-machining center. Because the flexible machining center is standardized, high volume

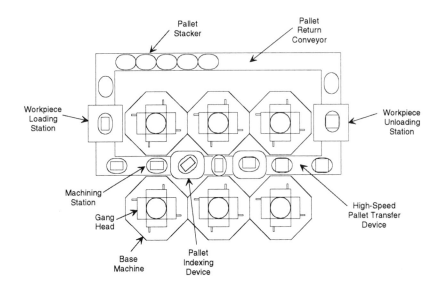

Figure 21.2 Flexible-machining center for finish machining engine block castings. Each machining station machines one orthogonal direction of the engine block.

production requirements are satisfied by duplicating the machining center a number of times. The elegance of this solution lies not only in the fact that new engine block designs can be introduced quickly, but also in the fact that ramp-up to a new design can be managed easily and economically. In the early stages of the new engine life cycle, only one or two machining centers are needed to produce the quantities required. As the new engine gains popularity and the engine it replaces phases out, more and more centers are shifted from the old engine to the new engine. In essence, the engine block design, production tooling, and production volume are all "decoupled" by this internal standardization solution.

21.5 STANDARDIZE COMMON PARTS

The goal in developing standardization approaches for unique parts is to make them as easy to design and introduce into production as possible. This is because a new and different unique part must usually be designed for each new product model or variant. For common parts, just the opposite is true. The goal in designing common parts is to avoid having to design them at all. This is done by creating a catalog of variants that meets all conceivable needs and that can be used interchangeably in all future new products. In creating the catalog of common parts, it is essential to avoid excessive part proliferation. One way for doing this is to use the S&R techniques discussed in Chapter 20.

To minimize information content, the catalog should be developed by (1) minimizing the number of different part categories (part families), (2) minimizing the number of variations within each category, and (3) minimizing the number of different design features within each variation. Once developed, the standard common parts should be used exclusively in new product designs. Also, where deemed both cost effective and beneficial, existing designs should be modified to accept the standardized design in place of special, one of a kind, versions of the component.

21.5.1 Evolutionary Approach

In general, two different approaches to creating internally standardized parts are possible: (1) evolutionary and (2) revolutionary. In the evolutionary approach, standardization is introduced gradually using internally standardized components in new designs and retrofitting existing products on a case by case basis depending on the benefits and costs involved. We refer to this as the S&R (standardization and rationalization) approach since it is an application of the S&R philosophy discussed in Chapter 20.

The S&R approach is implemented by deriving internal standard component designs from existing parts. Major changes to the overall part decomposition strategy (product architecture) and working principle employed are avoided to minimize the need to redesign existing products. The S&R approach can be implemented using the following step-by-step procedure.

1. Sort all designed parts manufactured or purchased by the company into unique parts and common parts.

For the unique parts, do the following:

2. Divide the unique parts into part families. Develop a composite part for each part family (see Fig. 18.4). Develop a standardization strategy appropriate for the design and production of each composite part.

In developing the composite parts, do the following where possible:

3. Standardize and rationalize features at all levels to reduce information content.
4. Standardize the manufacturing processes and tooling based on the composite part.

5. Standardize so that individual parts can be obtained by skipping some steps and features in the manufacturing process.
6. Develop a plan for implementing the standardization strategy for each unique composite part. Implement each strategy as resources and opportunities for change become available.

For the common parts, do the following:

7. Divide the common parts into part families. Develop a catalog of standard designs for each part family. In doing this, strive to minimize the number of variations on all levels to reduce information content.
8. Use the standardized common components exclusively in all new designs. Evaluate existing designs, especially those that are not likely to be redesigned any time in the near future, on a case-by-case basis. Where cost effective and beneficial, make minor design changes which will allow standard common components to be used in future production of existing products.

The evolutionary approach is most appropriate in situations where the overall product architecture is appropriate for meeting customer needs, a large number of existing product families and/or variants are in production, and time and resources available for implementing sweeping changes are limited. The main advantage of the evolutionary approach is that it can be developed and introduced gradually over time without the need for major redesign efforts and capital investment. At the same time, the evolutionary approach may limit some of the potential benefits available from internal standardization. It may also perpetuate costly aspects of the existing design.

21.5.2 Revolutionary Approaches

Unlike the evolutionary approach, revolutionary approaches to internal standardization generally require total redesign of the product. How the redesign is implemented depends on the particular type of products and parts involved. It also depends on the marketing and manufacturing strategies of the company. As a result, there are as many different revolutionary approaches as there are creative minds to invent them. In all cases, however, both the part decomposition (product architecture) and the parts themselves are likely to be totally different from the current product. The up side of adopting a revolutionary approach is that the benefits of standardization can be fully leveraged. On the down side, revolutionary approaches can involve significant risk and large capital investment since new production methods and practices will probably be required.

21.6 CREATE A "BUILDING BLOCK" DESIGN

In a "building block" approach, a set of carefully defined internal standard components are designed which can be combined in different ways to produce all anticipated variants of the final product. Building block designs typically involve a "revolutionary" redesign of the product since the design and method of manufacture of the building blocks is likely to be considerably different from the existing design. Also, to further leverage the advantages of the standardization, the final assembly method is usually totally revamped to facilitate flexible production of different variants in lot sizes of one or more with rapid change over from one variant to another.

Consider, for example, the design of automobile gages. Most automobiles utilize a variety of different gages such as the fuel, temperature, and oil pressure gage. Each of these gages must be designed to meet slightly different appearance and functional requirements according to the make and model of automobile it is used in. To avoid the enormous amount of information content (and cost) associated with creating and manufacturing an endless variety of unique gages, Nippondenso Company, Inc. (Ford Motor Company, 1985) created the building block design illustrated schematically in Fig. 21.3. The principle of producing various kinds of gages in this system is the selective matching of standardized parts. Mathematically it is possible to manufacture 288 different gages using the building block variants shown. With this design and a z-axis, top-down design for assembly, Nippondenso was able to produce every type of gage that was produced before the standardization system, all on one common assembly line. When first introduced, about 500,000 gages (1 shift), in 60 models, were produced every month, in a cycle time of one second. Also, the line could be reset in just one second to produce a different model. This was done up to 200 times a day.

Often, companies that are already producing an existing product family are in the best position to create a successful building block product design. This is because identifying the right building block groupings and variants typically requires extensive experience and insight. The goal is to minimize the number of groupings and the number of variants in each grouping while maintaining the ability to meet all anticipated design requirements for several product generations into the future. Another goal is to design the building blocks to be assembled in a way that allows flexible production of all product variants on one common automated assembly line. A general methodology that can be used is as follows:

1. **Analyze Existing Products:** Analyze the existing product family to identify the internal standardization opportunities that are available. Where appropriate, use group technology (GT) techniques (see

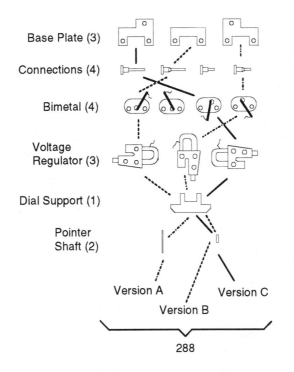

Base Plate (3)

Connections (4)

Bimetal (4)

Voltage
Regulator (3)

Dial Support (1)

Pointer
Shaft (2)

Version A   Version C

Version B

288

Figure 21.3 Selective matching of standardized building blocks to create up to 288 different automotive gages (Ford Motor Company, 1985).

Chapter 18) and other methods such as those discussed in Chapter 12 to understand underlying similarities across the different product families and product variants within each family.

2. **Create a Set of "Building Block" Components:** Decompose the product family into standardized building blocks (modules, sub-assemblies, components) that can be selectively combined in different ways to create all existing and conceivable future product variants.

3. **Develop a Coordinated Manufacturing Plan:** Develop a manufacturing approach that allows flexible production of any variant in lot sizes of one or more.

21.7 DESIGN WITH STANDARD FEATURES

Many of the standardization schemes discussed thus far have a down side in that they can significantly restrict the creative freedom the design team needs to optimally satisfy customer, business, and manufacturing requirements. The state of the art of solid modeling and modern parametric CAD/CAM systems offers the opportunity to extend the idea of combinative building blocks to the level of geometric form elements or features. The number of geometric features used by the design engineer is small compared to the number of possible designs that can be created by combining different versions of the geometric features. By standardizing geometric features instead of parts or subassemblies, many of the benefits of standardization can be reaped while avoiding the creative restrictions that can occur at higher levels of standardization.

21.7.1 Geometric Features

Geometric features are generic shapes or characteristics of various aspects of a component's geometry. Types of geometric features include (Shah and Mantyla, 1995):

- **Form Features:** features that describe portions of a part's nominal (or idealized) geometry.
- **Tolerance Features:** features that describe geometry variation from the nominal form.
- **Assembly Features:** features that describe the relationships between parts in a mechanical assembly.

The geometry of most components can be decomposed into features. To develop a library of features that can be used to design future versions of a component, the product must first be decomposed into component types. For example, an internal combustion engine is composed of several component types including pistons, connecting rods, crankshafts, camshafts, engine block, and so forth. Each component type will, in general, have it's own characteristic set of features.

In developing a feature set for each component type, it is important to recognize that features are also dependent on the design process used to create the component. That is, only features that are useful in component creation should be identified. Each feature should be a natural building block that fits the design logic used in creating the component. To be acceptable, a feature should satisfy the following basic requirements:

- It should serve a design function.
- It should recur in different versions of the component.
- It should be describable by a set of design parameters.
- It should be a part of the designer's vocabulary.

Consider shaft design, for example. A *shaft* is a rotating member, usually of circular cross section, used to transmit power and motion. The geometry of a shaft is generally that of a stepped cylinder where the steps consist of changes in diameter and grooves to provide location functions and to accept thrust loads. Steps and grooves are geometric features of the shaft because they meet the basic requirements:

- They serve a design function.
- They recur in different versions of the component.
- They can be described by a set of design parameters.
- They are a part of the designer's vocabulary.

Most components will have a variety of geometric features associated with different functional and manufacturing requirements. In addition to grooves and steps, a shaft has features such as keys and splines that transfer torque from one element to another on the shaft. Shah and Mantyla (1995) suggest the following systematic methodology for featurizing a component design:

1. Examine part drawings or CAD models of the component class that is to be featurized. Look at representative versions and several generations of the same part.
2. Identify regions of interest for a designer.
3. Identify macro shapes for different functional regions (mating region, containment region, reinforcement, etc.).
4. Decompose each region into the lowest level units that a designer would treat as single units; call these units "simple features." These units must satisfy the basic requirements for a feature given above.
5. Draw a generic sketch for each simple feature identified and label all its dimensions using the designer's terminology.
6. Determine if there are any "composite features," such as simple features arranged in a geometric pattern, or features with fixed interrelationships. Draw sketches of these composites and label their interrelationship parameters.

Once a set of features has been identified, the relationship of the features to the component's design process must be understood and characterized. The goal is to reconcile each feature with the design process so that it can be described in a way that makes it a useful building block for designing the component. Ultimately, each feature needs to be described using a uniform format in terms of the following (Shah and Mantyla, 1995):

- Generic shape (sketch)
- Dimension parameters
- Positioning reference entity (entities)
- Positioning method and positioning/orientation parameters
- Geometric constraints (intrinsic and extrinsic)
- Adjacency relationships with neighboring features.

21.7.2 Feature Standardization

Once an efficient and useful feature set for a particular component type has been developed, it can be standardized to reduce the information content of future component designs. One approach for doing this is to use the standardization and rationalization (S&R) procedure presented in Chapter 20. In standardizing the features for a particular component type, the method of manufacture that will be used should be considered. For example, if the component type is to be cast and then machined, the feature set should be standardized to facilitate and ease use of these processes.

The design with features approach is still in its infancy and additional research and experience is needed to fully understand the opportunities for information content reduction that are available at the feature level. It is clear however, that standardization coupled with guided common sense is applicable to features in many ways and it should be considered from the very beginning of a feature development program. Many of the strategies discussed in this chapter are also applicable and are recommended for consideration by those readers who are involved in feature development projects.

21.8 KEY TAKEAWAYS

- Information content of the manufacturing system is dramatically reduced when internal components have been standardized to simplify their design and manufacture and to minimize the proliferation of new part numbers within the firm.
- Components can be standardized in a variety of ways. For example,
    - The same subassembly or part can be used interchangeably in different products.
    - Different versions of the component can be produced on the same production line without the need for setup or time-consuming changeover.
    - Different versions of the product can be produced by assembling different combinations of standardized "building block" parts.
    - Different versions of the part can be designed by using different combinations of standard features.

22

Design Improvement Methods

22.1 INTRODUCTION

In practice, a great deal of product design effort is concerned not with the creation of revolutionary new products but with the evolutionary improvement of existing products. Even with revolutionary new products, an iterative redesign process is typically required to improve the product design and ready it for economical production (see Fig. 7.2). These redesigns or design modifications are usually aimed at improving product performance, reducing weight, reducing cost, improving reliability, enhancing appearance, improving manufacturability, improving manufactured quality, and so forth. A variety of engineering methods are available for assisting this improvement process. In this chapter, we present some of the more effective and useful of these.

22.2 GUIDED ITERATION

Many engineering design problems are solved in an iterative, guess-and-check fashion. That is, a design solution is first proposed, then analyzed and modified until it is "good enough" to be accepted. Dixon and Poli (1995) have formalized this process as a general-purpose problem solving approach, which they term *guided iteration*. The methodology is, in essence, a systematic application of the design-analyze-redesign strategy (Fig. 22.1). It involves four basic steps:

1. Formulate the design problem.
2. Generate one or more alternative solutions.
3. Evaluate the alternative solutions.
4. If none are acceptable, redesign, guided by the results of the evaluation.

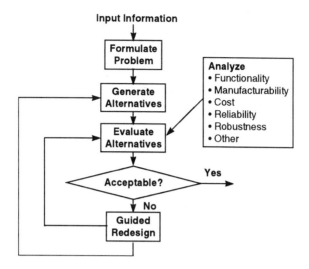

Figure 22.1 The guided iteration methodology (Dixon and Poli, 1995).

To illustrate the guided design process, suppose we wish to design a helical compression spring, having squared and ground ends (total number of turns = active turns + 2), to fit in a 50 mm diameter hole and to fit over a 25 mm diameter rod. The spring is to exert a force of 90 N when it is 87.5 mm long and a force of 630 N when it is 50 mm long. Loading is essentially static. Clash allowance should be >10%. The spring is to be made of music wire (ASTM A228) and is to be "set" after winding. A preferred wire diameter is to be used (Table 22.1) and the number of turns should be specified in half-turn increments.

Step 1: The problem is formulated in terms of nomenclature, parametric relationships, and design requirements as follows:

Nomenclature:

k = spring rate, D = mean coil diameter, d = wire diameter
N_a = number of active turns, F_S = force to compress spring solid
F_{max} = maximum operating force, L_S = solid spring length
L_f = free spring length, L_{min} = spring length at F_{max}
τ_s = shear stress when compressed solid, S_{ys} = shear yield strength
ca = clash allowance, G = shear modulus, FS = factor of safety

Parametric Relationships (Shigley and Mischke, 1989):

$$F_s = F_{max} + k(L_{min} - L_s), \qquad L_s = (N_a + 2)d$$

$$D = \left(\frac{d^4 G}{8k N_a}\right)^{1/3}, \quad L_f = L_s + \frac{F_s}{k}, \quad ca = \left[\left(\frac{(L_f - L_s)k}{F_{max}}\right) - 1\right]100\%$$

$$\tau_s = \frac{8 F_s D K_s}{\pi d^3}, \quad K_s = 1 + 0.5\left(\frac{d}{D}\right), \quad S_{ys} = \frac{927}{d^{0.163}} \text{ MPa}$$

Design Requirements:

$$k = \frac{\Delta F}{\Delta L} = \frac{630 - 90}{87.5 - 50} = 14.4 \text{ N/mm}, \quad F_{max} = 630 \text{ N @ } L_{min} = 50.0 \text{ mm}$$

$$FS = S_{ys}/\tau_s > 1.25, \quad OD = D + d < 50 \text{ mm}, \quad ID = D - d > 25 \text{ mm}, \quad ca \geq 10\%$$

Number of active turns, N_a, to be a whole or half number of turns.

Wire diameter, d, to be selected from Table 22.1.

Steps 2-4: An acceptable solution is iteratively evolved as follows:

1. As a starting point, we arbitrarily choose $d = 3$ mm and $N_a = 13$ turns.
2. Calculate the spring parameters to determine if the design meets the stipulated requirements. If all requirements are met, accept the design and terminate the process. If all requirements are not met, redesign based on insights gained and reanalyze. Results for several iterations are tabulated in Table 22.2.
3. In trial 1, the proposed design is unacceptable because $FS < 1.25$ and $ID < 25$ mm. Examination of the parametric relationships shows that increasing the wire diameter (d) will increase the factor of safety (FS). This provides the guidance required for the next iteration, trial 2. Now that the factor of safety is acceptable, the design relationships show that decreasing N_a will increase the spring ID. This provides the guidance required for trial 3. Using similar reasoning for each subsequent iteration, an acceptable design is identified after four tries.

In this simple example, only one design alternative is generated and evaluated at a time. In more complex problems, several alternatives may be generated. If we were considering the configuration design of the spring, for example, several alternative spring types such as torsional springs and belleville springs might be considered in addition to the helical coil spring. It

Table 22.1 Preferred wire diameters (mm)

1.00, 1.10, 1.20, 1.40, 1.60, 1.80, 2.00, 2.20, 2.50, 2.80, 3.00, 3.50, 4.00,
4.50, 5.00, 5.50, 6.00, 6.50, 7.00, 8.00, 9.00, 10.0, 11.0, 12.0, 13.0, 14.0

Table 22.1 Results of the Guided Iteration Process

Trial	d mm	N_a	FS	OD mm	ID mm	ca %	Insight guiding next iteration
1	3.00	13.0	1.00	19.2	13.2	11.4	Increase d
2	4.00	13.0	2.24	27.8	19.8	-22.6	Decrease N_a
3	4.00	8.0	1.21	32.0	24.0	22.6	Increase d
4	4.50	8.0	1.59	37.3	28.3	11.4	ACCEPTABLE

should be noted that the guided iteration method is used effectively at all stages of engineering design including conceptual, configuration, and parametric design. The goal of the methodology is to help insure acceptable design solutions. When employed with diligence and intelligence, it will produce satisfactory, and often excellent, results.

22.3 ANALYTICAL OPTIMIZATION

The goal of design optimization is to identify a design solution that satisfies all design requirements and is best in some sense. Best can be based on cost, weight, strength, capacity, and so forth. In most design situations, the design team is constantly striving to optimize the design. Often, this is done on a subjective basis using engineering experience and general principles. For example, by using the engine block to help support structural load, weight of a high performance motorcycle design is reduced. Compromise among several desirable characteristics is also often necessary. The low cost design and the high performance design are seldom the same.

When the problem of design can be expressed in explicit mathematical form, then analytical optimization becomes possible. In *analytical optimization*, the design objective (minimize cost, maximize energy absorption, etc.) as well as all of the design requirements or constraints are expressed mathematically in terms of design variables. Solution methods are then employed to determine the optimum numerical value for each design variable. When this is possible, analytical optimization can be an excellent design improvement tool for performing parametric design.

The analytical optimization process is similar to guided iteration except that the process is continued until the optimal design is identified. The general steps involved are as follows:

1. Specify the design variables.
2. Specify the objective function.
3. Specify the constraints.
4. Specify a solution procedure for determining the optimal design.

Design variables are numerical quantities for which values are to be chosen in producing the design. In the helical spring example of Section 22.2, k, d, N_a, and L_f are all design variables. Since different materials can be used, material is also a design variable. The objective function, which is expressed as a computable function of the design variables, is a measure of the design characteristic that is to be optimized. The task of the optimization is to find the set of design variable values that minimizes (or maximizes) the objective function.

Constraints are design requirements that must be satisfied in order to produce an acceptable design. Three different types of constraints are typically encountered. The first type arises due to performance or behavior requirements. The requirement that the spring material not yield when compressed solid is an example. The second type of constraint is one that restricts the range of design variables for reasons other than the direct consideration of performance. Space limitations such as the requirement that the helical coil spring fit in a 50 mm diameter hole is representative of this type of constraint. A third type of constraint occurs when the design variable is permitted to take only a discrete set of values. The preferred wire diameters given in Table 22.1 and the requirement that the number of turns be specified as whole or half-turns are examples.

There are many solution methods available depending on the nature of the problem. These include calculus-based methods, systematic search methods, and linear programming methods. Development and use of these methods are discussed in the many excellent texts and reference books on the subject. See for example Wilde (1978), Arora (1989), Paplambros and Wilde (1988), Haug and Arora (1979), and Kuester and Mize (1973).

Use of analytical optimization as a design improvement method can be illustrated by using it to select numerical values for the design variables in the helical compression spring design developed in Section 22.1. Assume that our objective in this exercise is to minimize the weight of the spring.

1. Based on the problem statement given in Section 22.1, the design variables are:

    Wire diameter: d (mm)
    Number of active turns: N_a

2. The objective function is:

$$W_a = \frac{\pi^2 \gamma}{8} \left(\frac{d^{10} G N_a^2}{k} \right) \to \min$$

where W_a is the weight of the active coils and γ the material's weight density.

3. The behavior, geometric, and discrete value constraints are:

$$k = 28.8 \text{ N/mm}$$

$$FS = \frac{S_{ys}}{\tau_s} > 1.25$$

$$OD = D + d < 50 \text{ mm}$$

$$ID = D - d > 25 \text{ mm}$$

$$ca > 10\%$$

$$d = \text{preferred size from Table 21.1}$$

$$N_a = \text{half or whole turns}$$

4. With this formulation, a number of optimization methods could be used to find the solution. Because there are only two design variables, it is a relatively simple matter to develop a systematic search method tailored to this problem. Looking at the objective function, we see that *Weight* $\to$ *min* when $d \to$ *small* and $N_a \to$ *small*. For small d, *ID* controls; for small N_a, shear stress controls. Based on these insights, we develop the following solution method:

    1) Select a discrete trial value for the wire diameter d from Table 22.1.
    2) Using the trial d, calculate the value of N_a that makes $FS = 1.25$
    3) Using the trial d and value for N_a from step 2, calculate the inside diameter (*ID*).
    4) If $ID < 25$ mm, increase d to the next larger size; if $ID > 25$ mm, decrease d to the next smaller size.
    5) Repeat steps 2, 3, and 4 until the value of d that satisfies all constraints and minimizes the weight of the spring is found.
    6) Increase N_a to next whole or half-turn to satisfy the discrete value constraint.

Table 22.3 Systematic search to find optimum spring design

Trial	d mm	N_a	FS	OD mm	ID mm	ca %	Weight N	Change
1	4.0	8.30	1.25	31.7	23.7	20.1	0.686	Increase d
2	4.5	6.12	1.25	40.4	31.4	30.7	0.829	Increase N_a
3	4.5	6.50	1.31	39.6	30.6	26.8	0.863	Optimum

Results of the search are tabulated in Table 22.3. Note that the discrete value constraint helps the process to converge quickly. Although this is a relatively simple example, it is representative of the type of parameter optimization encountered in many mechanical design situations. By formulating the problem correctly and developing a simple solution algorithm using a spreadsheet or other available software, analytical optimization can be a very effective method for making design decisions in these situations.

Because there are only two design variables involved, it is possible to plot the feasible design region graphically as shown in Fig. 22.2. Looking at this figure, we see that by decreasing the diameter of the rod over which the spring must fit, and relaxing the factor of safety and/or clash allowance requirements slightly, the feasible design region can be enlarged to allow a lower weight spring design. For example, if the rod diameter and factor of safety are reduced to 17 mm and 1.21, respectively, a spring weighing 0.530 N (d = 3.5 mm and N_a = 11.0 turns) would be feasible. This illustrates the idea stimulating aspect of design optimization that was discussed in Section 8.4.

22.4 TAGUCHI METHOD

Analytical optimization seeks to optimize the design with respect to performance and functionality. The design technique that has come to be known as the *Taguchi Method*, on the other hand, seeks to optimize the design with respect to robustness. A *robust design* is one that functions and performs as designed (performance and functionality are centered at design intent) with minimal variation due to hard-to-control manufacturing, environmental, use, and time related factors. The Taguchi method provides a systematic approach for maximizing robustness. The method gets its name from the efforts of Dr. Genichi Taguchi and the cost-saving approaches to quality pioneered in Japan.

Maximizing robustness requires a three-step optimization process: system design, parameter design, and tolerance design. System design involves the conceptual and configuration design of the product and the tentative selection of values for the design parameters. By selecting the right design concept, part configurations, materials, manufacturing processes, and so forth, the design is

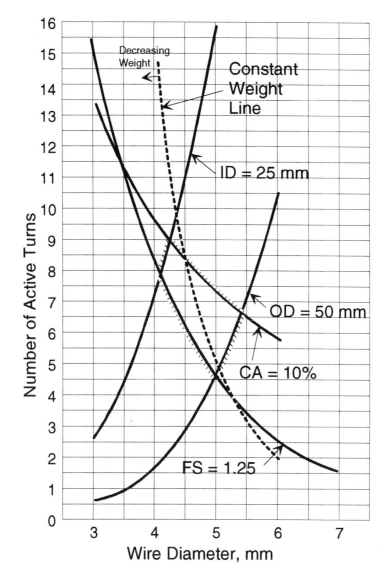

Figure 22.2 The feasible design region, which contains all acceptable combinations of wire diameter and active number of turns, is defined by the constraints as shown. Note that the constant weight contour (dashed line) shifts to the left for lower weight springs. Note also that the *ID* constraint moves up and to the left and the *FS* constraint moves down and to the left as these constraints are relaxed.

endowed with inherent robustness. This inherent robustness is then maximized in the parameter design step by selecting numerical values for the design parameters, which minimize sensitivity to variation of hard-to-control factors. If the reduced variation obtained in the parameter design step is not sufficient, tolerances on influential design parameters are tightened in the tolerance design step. Tolerance design usually means spending money on higher precision, better grade materials, and manufacturing process complexity. The great value of the Taguchi method is its focus on parameter design as a means for avoiding the need for tight tolerances by maximizing robustness.

The goal of parameter design is to systematically select values for controllable factors such that sensitivity to uncontrollable factors is minimized (see Fig. 17.6). Controllable factors include design parameters (e.g., dimensions, material properties, etc.) and other parameters such as manufacturing process settings. Uncontrollable or hard-to-control factors, called noise by Taguchi, include environmental factors, such as temperature and humidity, time and use factors, such as corrosion and wear, and manufacturing-related factors, such as part-to-part dimensional and material property variation.

The parameter design is performed using design of experiments (DOE) techniques. DOE provides the framework needed to define the scope of the investigation, set goals and formulate a hypothesis to evaluate, and analyze data. DOE also helps avoid the problems of one factor at a time testing. Answers are generated in an economical and efficient manner, they have meaning over a wide range of factor values, and they provide insight into interaction between factors.

Fractional factorial experiments (Fig. 22.3), called *orthogonal arrays*, are utilized to maximize generation of needed information while minimizing the number of experiments that must be run. The design of the orthogonal array depends on the number of parameters and the number of parameter levels to be studied (Fig. 22.4). Experiments are conducted at each combination of parameter values stipulated by the orthogonal array. Typically, a number of experiments must be performed at each parameter level combination to measure the effect of noise (hard-to-control variation) on the performance characteristic of interest. When it is possible to control the noise at predetermined levels, an outer array can be used to define the combination of noise levels to be used at each repetition. This often enables valid results to be obtained with a minimum number of repetitions.

Analysis of the experimental results is facilitated by calculating "signal-to-noise (S/N)" ratio values for each set of experiments. The S/N ratio is a figure of merit that combines the performance characteristic of interest and its variation into one measure. Taguchi has developed numerous S/N ratios for different applications (Fig. 22.5). S/N ratios are formulated so that robustness

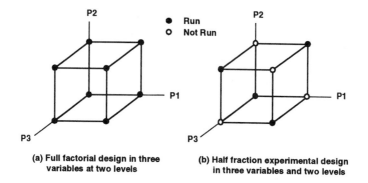

(a) Full factorial design in three
variables at two levels

(b) Half fraction experimental design
in three variables and two levels

Figure 22.3 Design of experiments generates information by testing at different combinations of factor values. (a) A full factorial design tests all combinations while (b) a fractional experiment tests a selected subset of combinations.

increases as the numerical values of the S/N ratio increases. To maximize robustness, the S/N values for each set of experiments are analyzed statistically using DOE techniques to identify the combination of parameter values that maximize the S/N ratio, and hence the robustness of the design.

The Taguchi method is implemented in six basic steps:

1. Determine the goal of the study.
2. Select the design parameters (control factors) and noise factors (hard-to-control factors) to be studied.
3. Design the experiment by selecting the number of levels (parameter values) to be used and deciding whether pure repetitions or an outer array will be used. Based on this, select the appropriate orthogonal array(s), S/N ratio, and numerical parameter values for each level.
4. Perform the experiments.
5. Calculate the S/N value for each experiment and analyze the results. Use insights gained to either (1) repeat steps 1 through 4 for improved results or (2) recommend optimum values for the design parameters.
6. Perform a confirming experiment using the optimal parameter values to validate the recommendations.

To illustrate this procedure in a very simplistic manner, let us reconsider the coil spring design discussed previously in Sections 22.2 and 22.3. Suppose we are concerned about the variation in spring rate (k) from spring-to-spring. The goal of the study is therefore to identify acceptable numerical values for d and N_a that minimizes this variation. Note that in this example, the design

L4 Array

Trial No.	Column No.		
	1	2	3
1	1	1	1
2	1	2	2
3	2	1	2
4	2	2	1

L8 Array

Trial No.	Column No.						
	1	2	3	4	5	6	7
1	1	1	1	1	1	1	1
2	1	1	1	2	2	2	2
3	1	2	2	1	1	2	2
4	1	2	2	2	2	1	1
5	2	1	2	1	2	1	2
6	2	1	2	2	1	2	1
7	2	2	1	1	2	2	1
8	2	2	1	2	1	1	2

(a) Two Level Orthogonal Arrays

L9 Array

Trial No.	Column No.			
	1	2	3	4
1	1	1	1	1
2	1	2	2	2
3	1	3	3	3
4	2	1	2	3
5	2	2	3	1
6	2	3	1	2
7	3	1	3	2
8	3	2	1	3
9	3	3	2	1

(b) Three Level Orthogonal Array

Figure 22.4 Example orthogonal arrays (Ross, 1988).

parameters (d and N_a) are known from the problem statement. In a more open-ended design situation, the important design variables may not be immediately obvious and some repetitions of steps 2, 3, and 4 may be required to identify them.

The spring rate is likely to vary from spring to spring because of variation in many of the manufactured features of the spring. For example, there will be some dimensional variation of the spring wire diameter (d). Also, because of spring-back and other hard to control aspects of the manufacturing process, the free length of the spring, actual number of turns, mean coil diameter (D), and geometry of the squared and ground ends are likely to be slightly different for each spring produced. Because all of these noise factors are relatively hard to control, even under very controlled test conditions, we decide not to select any specific noise factors for study.

The orthogonal array specifies the combinations of parameter values to be used. It is selected based on the number of design variables involved and the number of values for each variable (levels) to be used. For this example, we decide to use two levels to minimize the amount of testing required. The simplest orthogonal array for a two-factor (d and N_a), two level experiment is

Type NB: Nominal is Better (Dimensions, Voltage, etc.)

$$S/N_{NB1} = -10\log V_e$$

$$S/N_{NB2} = +10\log\left(\frac{V_m - V_e}{r V_e}\right)$$

Where $\qquad V_m = \dfrac{\left(\sum y_i\right)^2}{r} \qquad V_e = \dfrac{\sum y_i^2 - \left(\sum y_i\right)^2 / r}{r - 1}$

y_i = an observation r = the number of observations

Type LB: Lower is Better (Noise, Stress, etc.)

$$S/N_{LB} = -10\log\left(\frac{1}{r}\left(\sum y_i^2\right)\right)$$

Type HB: Higher is Better (Strength, Power, etc.)

$$S/N_{HB} = -10\log\left(\frac{1}{r}\left(\sum \frac{1}{y_i^2}\right)\right)$$

Figure 22.5 Some useful *signal-to-noise* (*S/N*) ratios (Ross, 1988).

the L4 array (Fig. 22.4). An outer array is used when the level of selected noise factors is to be controlled. Since we decided that none of the noise factors could be realistically controlled, an outer array is not specified.

Selecting appropriate parameter values for each level typically requires engineering judgment and insight. For this example, we gain the insight needed by viewing Fig. 22.2, which shows that the feasible design region is bounded by wire diameters (d) of 4.0 and 5.5 mm and active turns (N_a) of 4.0 and 9.0. By selecting these values for the lower and upper levels of d and N_a, respectively, we ensure that the experimental investigation will span all feasible numerical values for the design parameters.

The choice of *S/N* ratio depends on the goal of the study and type of characteristic involved. For example, if reducing stress in a load-bearing structure were the goal, lower stress would be better. If, on the other hand, increased load carrying capacity of a glued joint is the goal, then higher joint strength would be better. In still other cases, such as dimensional variation, nominal is better. In the case of the spring rate, we decide that nominal is better. Also, since we are adjusting the spring parameters to produce a desired spring rate (14.4 N/mm), our concern is with variation of this characteristic. Accordingly, we select *S/N_NB1* from Fig. 22.5.

To conduct the experiment, four sets of 25 springs are wound, with each set having a different combination of numerical values for d and N_a and all having the required force-deflection characteristics specified in Section 22.2. The sample size of 25 is chosen based on the cost and time involved. In general, confidence in the results will increase as the sample size increases, so sample size should be as large as practical when no outer array is used. After winding the springs, the spring rate (k) of each spring is measured, recorded, and used to calculate an S/N ratio for each parameter combination. Results are shown in Fig. 22.6a.

The experimental results are analyzed by plotting average S/N ratio verses wire diameter (d) and active number of turns (N_a) as shown in Fig. 22.6b. The average S/N ratio is the average of all the S/N values computed at each parameter level. For example, the average S/N for $d = 4.0$ mm is calculated as

$$Avg(S/N) = \frac{6.990 + 10.432}{2} = 8.711$$

Taguchi has formulated the S/N ratios so that robustness is maximized when $S/N \rightarrow \max$. From Fig. 22.6b, we see that S/N increases with wire diameter (d) and the active number of turns (N_a). Therefore, to minimize spring-to-spring variation in spring rate, we would like to use the largest wire diameter and number of active turns possible. Looking at Fig. 22.2, the largest wire diameter that can be used is 5 mm and for this wire diameter, the largest feasible value for N_a is 7. We therefore set $d = 5$ and $N_a = 7$ to maximize robustness with respect to spring-to-spring variation in spring rate (k). To validate the design, we manufacture 25 springs to this specification ($d = 5$ mm, $N_a = 7$ turns) and, after measuring k for each spring, find the standard deviation of the spring rate to be 0.31 N/mm.

The Taguchi method has been widely used to improve the quality and robustness of products and systems. Some recent books on the subject include Ross (1988), Bendell, Disney, and Pridmore (1989), Lochner and Matar (1990), Pradke (1989), and Peace (1993).

22.5 PROBABILISTIC DESIGN

Material properties, component dimensions, externally applied loads, and so forth are multivalued; that is, they vary from lot to lot, product to product, and application to application. The *probabilistic design* approach seeks to incorporate real-world randomness and variation into the design decision making process by treating design parameters and characteristics statistically rather than deterministically. Probabilistic design methods help improve parameter design decisions in several ways. Because variation is considered in

Trial No.	d	N_a	S/N
1	4.0	4.0	6.990
2	4.0	9.0	10.432
3	5.5	4.0	7.469
4	5.5	9.0	11.245

(a) Layout of Experiment and Results

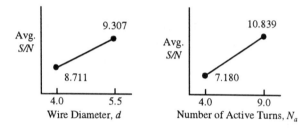

(b) Analysis of Results

Figure 22.6 Use of Taguchi method to optimize robustness of the coil spring design with respect to variation of spring rate.

the analysis, probabilistic design allows reliability goals to be determined by design. This not only helps to ensure known levels of performance and safety, but also helps avoid costly and time-consuming testing and redesign as well as added material and processing cost that may result from over design. Another important use of probabilistic design techniques is in the analysis and reduction of undesirable consequences produced by tolerance stack-up in complex assemblies. Also, when it is possible to model a design problem mathematically, probabilistic design can be used to analyze variation and optimize robustness.

Probabilistic design requires that the design variables be described in terms of their means, standard deviations, and distributions. When the problem of design is very complex or when different probability distributions are associated with the various parameters involved, it is often best to employ computer-based simulation techniques. In these techniques, values for each variable are selected at random from their respective distributions and combined according to a mathematical model of the design problem. Commercially available software using this and other approaches is available for applications such as tolerance stack-up analysis.

When all of the design parameters are normally distributed, then it is possible to perform probabilistic design analytically. Doing this, however, requires knowledge of how the error or uncertainty propagates through the design calculations. Suppose for example that we wish to add two random variables to form a third random variable. To do this, we must first make the assumption that all stochastic design parameters are independent, random, and normally distributed. Each design variable can then be characterized by specifying a mean and standard deviation. To do this, we adopt the following nomenclature:

- The *mean* of a stochastic design variable x is designated as $\bar{x}$
- The *standard deviation* is designated as $\hat{\sigma}_x$

- The *coefficient of variation* is defined as $C_x = \dfrac{\hat{\sigma}_x}{\bar{x}}$

With these definitions, the mean for the sum $z = x + y$ would be

$$\bar{z} = \bar{x} + \bar{y}$$

and the standard deviation, which follows the Pythagorean theorem, would be

$$\hat{\sigma}_z = \sqrt{\hat{\sigma}_x^2 + \hat{\sigma}_y^2}$$

Similar relations have been developed for a variety of functions, some of the more useful of which are given in Table 22.4.

22.5.1 Reliability Design

To illustrate how the probabilistic design approach can be used to design for reliability, consider the selection of an appropriate factor of safety to guard against uncertainties associated with material properties, magnitude of external loading, part-to-part dimensional variation, and so forth. Let P_f designate the theoretical critical failure value and P_w the safe working value associated with a particular design situation and failure mode. Then the factor of safety (*FS*) guarding against failure is,

$$FS = \frac{P_f}{P_w} \tag{22.1}$$

Table 22.4 Means and coefficients of variation for simple operations (Shigley and Mischke, 1989)

Function	Mean Value	Coefficient of Variation
$x + y$ $_X$	$\bar{x} + \bar{y}$	$\dfrac{\sqrt{\hat{\sigma}_x^2 + \hat{\sigma}_y^2}}{\bar{x} + \bar{y}}$
$x - y$	$\bar{x} - \bar{y}$	$\dfrac{\sqrt{\hat{\sigma}_x^2 + \hat{\sigma}_y^2}}{\bar{x} - \bar{y}}$
xy	$\bar{x}\,\bar{y}$	$\sqrt{C_x^2 + C_y^2}$
$\dfrac{x}{y}$	$\dfrac{\bar{x}}{\bar{y}}$	$\sqrt{C_x^2 + C_y^2}$
$\dfrac{1}{x}$	$\dfrac{1}{\bar{x}}$	C_x
x^2	$\bar{x}^2$	$2C_x$
x^3	$\bar{x}^3$	$3C_x$
x^4	$\bar{x}^4$	$4C_x$

The magnitude selected for the factor of safety (*FS*) typically depends on the uncertainties involved and on experience with similar situations.

To use probabilistic design to help select an appropriate factor of safety, we assume the critical value (P_f) and the working value (P_w) are both normally distributed stochastic variables. Failure will occur when $P_f - P_w \leq 0$. This condition occurs in the region of interference between the critical and working value distributions as shown in Fig. 22.7a. Since both the critical value and the working value are assumed to be normally distributed, a new variate X can be defined as

$$X = P_f - P_w \tag{22.2}$$

for which, from Table 22.4,

$$\bar{X} = \bar{P}_f - \bar{P}_w \tag{22.3}$$

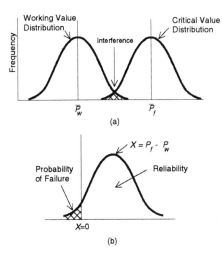

Figure 22.7 Interference theory of reliability prediction: (a) failure occurs in the region of overlap between the two distributions (i.e., $X = P_f - P_w \leq 0$); (b) $X = P_f - P_w$ is normally distributed.

and

$$\hat{\sigma}_X = \sqrt{\hat{\sigma}_f^2 + \hat{\sigma}_w^2}$$ (22.4)

The distribution of X is shown in Fig. 22.7b.

To estimate the percentage of parts that are likely to fail (i.e., $P(X \leq 0)$), we transform X into the standard normal distribution using

$$Z_\alpha = \left| \frac{X - \overline{X}}{\hat{\sigma}_X} \right|$$ (22.5)

Combining Eqs. (22.3), (22.4), and (22.5) and setting $X = 0$ gives the value of Z_α (designated Z_0) corresponding to the probability of failure

$$(Z_\alpha)_{failure} = Z_0 = \left| \frac{-(\overline{P}_f - \overline{P}_w)}{\sqrt{\hat{\sigma}_f^2 + \hat{\sigma}_w^2}} \right|$$ (22.6)

Table 22.5 Reliability factor

% Failure	% Reliability	Z_0
50.0	50.0	0
10.0	90.0	1.288
5.0	95.0	1.645
1.0	99.0	2.326
0.1	99.9	3.091
0.01	99.99	3.719
0.001	99.999	4.272

When all the terms on the right side of Eq. (22.6) are known, Z_0 can be calculated and the probability of failure determined from Table A1 in the Appendix. For convenience, we list selected values of Z_0 and the corresponding probabilities of failure in Table 22.5. Also tabulated is the reliability, where reliability (R) is defined as

$$R = P(X > 0) = 1 - P(X \le 0) \tag{22.7}$$

We will refer to Z_0 as the "reliability factor" since its value is determined from Table 22.5 (or Table A.1) based on a desired reliability. Equation (22.6) can be rearranged to provide the following expression for factor of safety (FS) as a function of the reliability factor

$$FS = \frac{\overline{P_f}}{\overline{P_w}} = \frac{1 + Z_0 \sqrt{C_f^2 + C_w^2(1 - C_f^2 Z_0^2)}}{1 - C_f^2 Z_0^2} \tag{22.8}$$

To illustrate use of Eq. (22.8) and Tables 22.4 and 22.5, suppose the coil spring discussed previously in Sections 22.2 through 22.4 is to be designed such that only about 1 spring in every 1000 springs will suffer permanent deformation when compressed solid. Suppose also that a large database of test data gathered over many years indicates that $C_f = 0.08$. Does the minimum weight spring specified in Section 22.3 meet this requirement?

From Section 22.2, the shear stress when the spring is compressed solid is

$$\tau_s = \frac{8 F_s D}{\pi d^3} \left(1 + \frac{0.5d}{D}\right)$$

Using the relationships in Table 22.4 gives

$$C_w = \frac{\hat{\sigma}_{\tau_s}}{\bar{\tau}_s} = \sqrt{C_D^2 + (C_d^2 + C_D^2) + (3C_d)^2} = \sqrt{2C_D^2 + 10C_d^2}$$

Values for C_D and C_d can be determined from helical spring manufacturer handbook data. For the minimum weight spring specified in Section 22.3 ($d = 4.5$ mm, $N_a = 6.5$ turns), handbook data gives $C_D = 0.004$ and $C_d = 0.00225$ for which we calculate

$$C_w = \sqrt{2(0.004)^2 + 10(0.00225)^2} = 0.0091$$

The 1 in 1000 failure rate is equivalent to 0.1% probability of failure or 99.9% reliability. From Table 22.5, $Z_0 = 3.091$. Substituting into Eq. (22.8) gives

$$[FS]_{R=99.9\%} = \frac{1 + 3.091\sqrt{(0.08)^2 + (0.0091)^2[1 - (0.08)^2(3.091)^2]}}{1 - (0.08)^2(3.091)^2} = 1.33$$

The actual factor of safety for the design of Section 22.3 ($d = 4.5$ mm, $N_a = 6.5$ turns), calculated using the relationships of Section 22.2 is

$$[FS]_{Calculated} = \frac{S_{ys}}{\tau_s} = 1.31$$

Since the factors of safety are about the same, we can conclude that about 1 in every 1000 of these springs are likely to yield when compressed solid.

It is important to note that probabilistic design depends on the availability of valid stochastic data and should be used with great care and judgment. For instance, the results of the above example could be quite misleading if the material strength properties were based on only a few tests or if the tested material is substantially different from that used in the actual part.

22.5.2 Analytical Robust Design

In Section 22.4, we used the Taguchi method to specify the most robust feasible spring design with respect to variability of the spring rate (k). Because the spring design relationships are modeled mathematically, it is also possible to use probabilistic design methods to investigate robustness. For example, the variability of k can be estimated using the relationships given in Table 22.4 and the following expression for k, which is derived from the design relationships given in Section 22.2:

$$k = \frac{d^4 G}{8D^3 N_a}$$

From Table 22.4, we have

$$\hat{\sigma}_k = \bar{k}\, C_k = \bar{k}\, \sqrt{(4C_d)^2 + C_G^2 + (3C_D)^2 + C_{N_a}^2}$$

Using this equation, in conjunction with handbook data, results from previous examples, and some assumptions (e.g., $C_G = 0.008$), we list and/or compute the probabilistic data shown in Table 22.6 for the various spring design alternatives developed in Sections 22.3 and 22.4.

Looking at these results, we see that the tendency toward less variation in k with increasing d and N_a predicted by the Taguchi method is confirmed. However, it is also seen that the maximum variation in k, which is estimated as plus and minus three times the standard deviation, is relatively small over the range of parameter values investigated. On the other hand, the penalty in increased spring weight is high for the maximum robustness design. Based on these insights, it appears that the minimum weight design is the best overall design. We also see that a significant further weight reduction is possible with little penalty in robustness by relaxing the ID and FS constraints as discussed in Section 22.3.

As a final comment, it is interesting to note that the standard deviation for the maximum robustness design ($d = 5.0$ mm, $N_a = 7.0$ turns) predicted by probabilistic design is 0.2538 N/mm which is less than the 0.031 N/mm found experimentally in Section 22.4. In particular, the experimental results appear to indicate a higher sensitivity to N_a than that predicted by the probabilistic analysis. One explanation might be that the prototype springs used in the test were not accurately or consistently made. Alternatively, it may be that one or more important factors are missing from the probabilistic analysis or the coefficient of variation data may be incorrect for this application. These inconsistencies emphasize the importance of good data and engineering judgment when using probabilistic methods to predict reliability or robustness.

Shigley and Mischke (1989) present a wide variety of probabilistic design applications in their *Mechanical Engineering Design* textbook. In particular, Chapter 4 of this reference presents an excellent overview of statistical considerations in design. For an in-depth discussion of all aspects of probabilistic design, see the text by Haugen (1980).

Table 22.6 Expected variation in spring rate (k) and weight calculated using probabilistic methods for coil springs specified in Sections 22.3 and 22.4.

Design	d	N_a	C_d	C_D	C_{Na}	Expected Variation of k ($\pm 3\hat{\sigma}$)	Weight
Minimum Weight Design from Section 22.3	4.5	6.5	0.00225	0.00399	0.00769	$\pm$0.81 N/mm	0.863 N
Maximum Robustness Design from Section 22.4	5.0	7.0	0.00200	0.00382	0.00714	$\pm$0.76 N/mm	1.288 N
Minimum Weight Redesign from Section 22.3	3.5	11.0	0.00290	0.00411	0.0045	$\pm$ 0.83 N/mm	0.530 N

22.6 VALUE ENGINEERING

Value engineering provides a systematic approach for design improvement. Typically, a multi-discipline team analyzes the functions provided by the design and the cost of each function. Based on results of the analysis, creative ways are sought to eliminate waste and unneeded function and to achieve required functions at the lowest possible cost. In value engineering, value is defined as

$$Value = \frac{Function}{Cost} \tag{22.9}$$

Since cost is a measure of effort, the value of a product or component using this definition is seen to be simply the ratio of output (function or performance) to input (cost) commonly used in engineering studies. In a complicated product design or system, every component contributes both to the cost and the function of the entire system. The ratio of function to cost of each functions indicates the relative value of functions performed by the product. Obtaining the maximum functionality per unit cost is the basic objective of the value engineering approach.

In general, the term "value engineering" applies more appropriately to the design of new products or the major redesign of an existing product. When emphasis of the redesign effort is on reducing costs of an existing design, the method is often referred to as *value analysis*. The focus of value analysis is mainly on the detailed design of components. Can a lower cost material and/or manufacturing process be used? Can the component shape or wall thickness or detail geometry be modified to reduce cycle time or increase yield? Is the specified tolerance and surface finish needed? Is a particular appearance feature worth the extra cost or could it be provided in a less costly but equally effective fashion? Value analysis relies heavily on the availability of accurate cost data.

Value engineering, on the other hand, is more of an umbrella methodology or generalized approach that is focused on the entire design process. For example, many of the structured methods discussed in this book are utilized as part of the value engineering approach. No matter what the level at which it is applied, however, the focus is always on function, cost, and the value formula (Eq. 22.8). How can value be added? Can the function be increased? The cost lowered? Can we do both?

For any expenditure or cost, two kinds of value are received: use (functional) value and esteem (prestige) value. Use value reflects the properties or qualities of a product or system that accomplishes the intended work or service. To achieve maximum use value is to achieve the lowest possible cost in providing the performance function. Esteem value is composed of properties, features, or attractiveness that makes ownership of the product desirable. To achieve maximum esteem value is to achieve the lowest possible cost in providing the necessary appearance, attractiveness, and features the customer wants. Examples of prestige items include surface finish, streamlining, packaging, decorative trim, ornamentation, attachments, special features, adjustments, and so forth. In addition to the two kinds of value received, additional costs are incurred due to unnecessary aspects of the design. Termed waste, these are features or properties of the design that provide neither use value or esteem value.

22.6.1 Value Analysis

Value analysis is generally performed in two phases, the analytical phase and the creative phase. In the analytical phase, the use value and esteem value offered by the product are systematically investigated by a team made up of experts representing all relevant components of the manufacturing enterprise. Findings generated in the analytical phase are then used by the team in the creative phase to define innovative design solutions which maintain the desired balance between use and esteem value, maximize these values by providing required function for the lowest cost, and eliminate identified waste or unneeded features. Some steps in a typical value analysis are as follows:

1. Identify the basic and secondary function or functions that the product, subassembly, or component is intended to perform. Basic functions relate to the specific work that is to be performed. Secondary functions

are subordinate functions that are performed in providing the basic functions.

2. Determine the value of all functions performed. This step is usually carried out in tabular form with each function forming a separate row. Percent of total performance contributed by each function is estimated and its cost as a percent of the cost of the product or subassembly is determined. Value is calculated as the ratio of percent performance to percent of cost. Insight into what aspects of the design constitute waste and should be eliminated and/or where improvements are needed is obtained by comparing the value calculated for each function.

3. Search for ways of reducing cost without reducing value, or of adding value without adding cost.

    •   Question the need for stated specifications and other design requirements. Compare the cost required to meet specifications with their value.

    •   Make a material analysis to determine the need for material properties used. Consider material substitutions and/or alternative manufacturing methods.

    •   Make a design analysis to determine alternative ways of performing the functions or eliminating certain functions.

    •   Question the value of high cost components or features. Consider modification or elimination of these features or the substitution of lower cost alternatives.

    •   Pay particular attention to components used in large number since small individual savings may add up to large overall savings.

    •   Find the lowest cost means to satisfy the function. This usually represents fairly decent value because of least cost. Compare this cost with the actual cost. If there is a large difference, then look for ways to reduce cost, even if the lowest cost version is not a viable option.

    •   Add value without increasing cost by improving the ease of use or user friendliness and by improving the ease of manufacture.

4. Evaluate, winnow, refine, and select improvements.

5. Redesign the product, subassembly, or component to incorporate the selected improvements.

22.6.2 Example

To illustrate value analysis, consider the airflow control " shutter" shown in Fig. 22.8. This device is used to control high-volume, low-pressure airflow through equipment such as heat exchangers or radiators, ventilation ducts, and other similar applications. The shutter controls airflow by rotating a series of

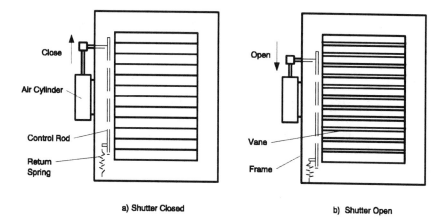

a) Shutter Closed   b) Shutter Open

Figure 22.8 The "shutter" device controls airflow by rotating vanes to open and close the the opening through which the airflow passes.

vanes in a manner similar to a "venetian blind" (Fig. 22.8). In the "closed" position, the shutter vanes are rotated to form a solid, relatively airtight front. Extruded rubber lip-seals, mounted in each vane, block airflow between the vanes when in the fully closed position. In the open position, the vanes are rotated to provide maximum space for air to flow between the vanes. The vanes are rotated using a linear pneumatic actuator and spring arrangement (Fig. 22.9). The vanes are mounted in the frame and connected together by the control rod to form a parallelogram linkage. Rotation of the vanes to the fully open position is produced by translation of the control rod as air pressure is applied to the air cylinder. Note that the shutter is held in the "open" position by a compression spring to provide "fail-safe" operation. Note also that many of the components used in the assembly "fasten" and/or "retain" other components and do not contribute directly to controlling the airflow (Fig. 22.10). Our goal in this example is to seek ways for reducing the cost of the existing shutter design using the value analysis method.

In the analytical phase, the team seeks to identify the basic and secondary functions performed by the shutter. Many techniques for doing this are discussed in the literature. See for example Fowler (1990) and Fallon (1980). In the case of the airflow control shutter, the team develops a functional decomposition using the approach discussed in Section 9.2.1 (Fig. 22.11). Once an understanding of the functions is developed, the team allocates a portion of the cost of each product component to the various functions (Fig. 22.12), estimates the percent of total product performance contributed by each function, and calculates the value of each function (Fig. 22.13).

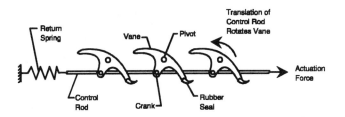

Figure 22.9 The shutter vanes are rotated by translation of the control rod.

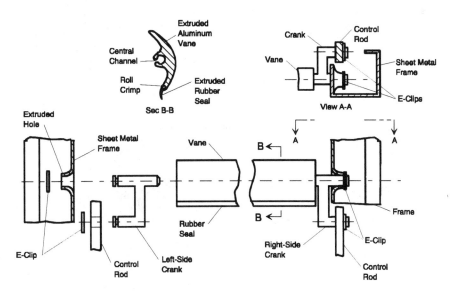

Figure 22.10 Details of the shutter construction.

The value analysis worksheet (Fig. 22.13) provides the insight needed to improve the design by eliminating waste and cost while maintaining or increasing function. For example, we see from Fig. 22.13 that the functions "provide décor" and "fasten/retain parts" have no value. Therefore, anything that can be done to eliminate these functions or reduce their cost will improve the value of the shutter design. Similarly, design changes that reduce the cost of "contain assembly" and "seal air" will increase value of the design. On the other hand, efforts to improve the "control opening," "transmit force," and "develop force" functions will have little impact since the value of these functions is already relatively high.

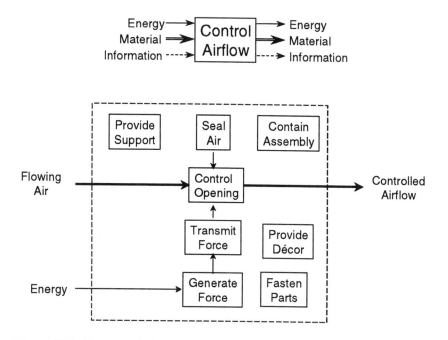

Figure 22.11 Functional decomposition of the airflow control shutter. Note that the basic function of the device is to "control opening." Performing this function, however, requires several secondary functions as shown.

The insights gained in the "analytical" phase guide the team in the "create" phase. For example, one way to reduce cost and increase value is to integrate the shutter frame members. Another way to improve the design is to eliminate various fasteners and to design components for ease of assembly. Also, the need to paint the device can be eliminated by substituting precoated steel or plastic for the frame members. Evaluating and selecting the design changes that are to be implemented ultimately depends on the tradeoffs between product cost, product performance, development cost, and development time.

Many of the methods and practices discussed elsewhere in this book have direct application in the value engineering method. Value engineering is unique, however, in its focus on function and on the cost of providing function. Because of its reliance on cost, value engineering can be difficult to use in the very early stages of design. On the other hand, it is an excellent starting point for the redesign of an existing product.

Component	Cost	FUNCTIONS (Verb-Noun)							
		Control Open-ing	Pro-vide Sup-port	Con-tain Ass'y	Seal Air	Trans-mit Force	Gener-ate Force	Pro-vide Decor	Fas-ten Parts
Frame Screw	0.31								0.31
Top Frame	1.59		0.64	0.64	0.24			0.08	
Bot Frame	2.22		0.89	0.84	0.33			0.11	
RS Frame	1.16		0.46	0.46	0.17			0.06	
LS Frame	0.9.3		0.37	0.37	0.14			0.05	
Actuator Bolt	0.08								0.08
Actuator	4.27	2.13					2.14		
Act. E-Clip	0.02								0.02
Act. Spacer	0.10					0.10			
Cont Bar Bkt	0.53					0.53			
Cont Bkt Scr.	0.04					0.02			0.02
Control Bar	1.78	0.89				0.53			0.36
Ret Spring	0.40	0.32					0.08		
Spring Mnt	0.07					0.05			0.02
Crank E-Clip	0.25								0.25
Crank	3.52					1.76			1.06
Vane	14.00	8.40				2.10		0.70	2.80
Vane Seal	1.66				1.56	0.10			
TOTAL	32.93	12.44	2.36	2.36	2.44	5.19	2.22	1.00	4.92

Figure 22.12 Cost-Function Worksheet. The team allocates a portion of component cost to each function.

Function	Estimated % Performance Contribution	Cost, $ (from Fig. 22.12)	% Cost	Value of Function
Control Opening	50	12.44	38	1.32
Provide Support	8	2.36	7	1.14
Contain Assembly	6	2.36	7	0.86
Seal Air	6	2.44	7	0.86
Transmit Force	20	5.19	16	1.25
Develop Force	10	2.22	7	1.43
Provide Decor	0	1.00	3	0
Fasten Parts	0	4.92	15	0
TOTAL	100	32.93	100	

Figure 23.13 Value Analysis Worksheet. The team estimates the percent performance contribution of each function and then calculates the percent cost and value of each function.

22.7 KEY TAKEAWAYS

- Searching for optimum conditions using nonsystematic trial and error methods can be time consuming and unproductive.
- Results obtained by using systematic design improvement methods are typically far better that those achieved by unstructured trial and error approaches.
- Guided iteration and analytical optimization methods provide the structure required to efficiently converge to the best design.
- The Taguchi Method provides a structured approach for systematically reducing hard-to-control variation. The design is improved by understanding the relationship between variability, the parameter values, and the tolerances on these values.
- Probabilistic design allows variation to be considered analytically provided the design problem can be modeled mathematically and that reliable statistical data are available for the key parameters involved.
- Value engineering methods provide a structured approach for investigating product function and the balance between performance, esthetics, and cost.
- Design improvement methods can be used at all stages and levels of design.

23

Failure Mode and Effects Analysis

23.1 INTRODUCTION

Failure mode and effects analysis (FMEA) is the name given to a group of activities performed to ensure that:

1. All that could potentially go wrong with a product has been recognized.
2. Corrective actions are taken to prevent potential failures from occurring.

Experienced design and manufacturing engineers instinctively analyze their designs and processes for potential failures. However, this type of improvised analysis is typically unstructured and undisciplined and is largely dependent on the personalities and experience of the individuals involved. FMEA, on the other hand, seeks to formalize and systematize the process to ensure that a new design is not released without being subjected to a formal review and correction procedure. FMEA imposes discipline by requiring formal documentation and rigorous procedures. Hence, successful use of FMEA generally requires that it be a planned step in the firm's product design process.

The primary goal of FMEA is to avoid surprises and unnecessary quality risk by preventing undiscovered design flaws and problems from reaching the customer. It does this by ensuring that all conceivable failure modes and their effects on operational success of the product have been considered. By listing potential failures and identifying the relative importance of their effects, it also helps to provide a basis for establishing corrective action priorities. A formal FMEA process also helps the design team in a variety of other ways:

- It assists in selecting design alternatives with high reliability and high safety potential during the early conceptual phases of product design.
- It provides a basis for product testing during development and final validation of the design.
- It helps the team develop early criteria for manufacturing, assembly, shipping, and service.
- It provides historical documentation for future reference to aid in the analysis of field failures and consideration of design changes.
- It helps ensure that the team presents defect prevention efforts to management during design reviews.

23.1.1 Types of FMEA

There are three general types of FMEAs: design, process, and service. The design FMEA generally focuses on the components, subsystems, and main systems of the product. The goal is to minimize failure effects on the system or product and to maximize system or product quality and reliability. The process FMEA is concerned with the method of product manufacture. It includes consideration of machines, tools, work stations, production lines, manufacturing processes, gauges, operator training, and so forth. The goal of the process FMEA is to minimize production process failure effects on the system or product, to maximize product quality and reliability, and to maximize manufacturing productivity. The service FMEA considers the tools, manuals, and procedures used in servicing and maintaining the system or product. The goal of the service FMEA is to minimize service failure effects on the system or product and to maximize customer satisfaction with the system or product. In keeping with the general subject matter of this book, our primary focus in this chapter is on the design FMEA.

23.1.2 FMEA Team

When performing an FMEA, a team approach is essential to insure that all needed product knowledge is available for brainstorming possible failures, effects, and corrective actions. For example, interaction with manufacturing and/or process engineering while conducting a design FMEA is important to ensure that the design can be manufactured per design specifications. Similarly, interaction with design engineering while conducting a process or assembly FMEA helps insure that design-related issues and causes are properly addressed. Most importantly, group consensus is needed to identify the high-risk areas that must be addressed to assure a high quality and reliable product.

Disciplines that should be represented on the design FMEA team include:

- Design Engineering

- Test Development Engineering
- Reliability Engineering
- Materials Engineering
- Field Service
- Manufacturing Engineering

23.2 METHODOLOGY

The FMEA process involves three main activities:

1. **Understand:** Evaluate the component, subsystem, or system to understand its intended function, the ways in which it could potentially fail to perform its function, and the safeguards that act to prevent these failures from occurring.
2. **Assess:** Estimate the risk associated with each potential failure to provide a basis for ranking the seriousness of each failure and prioritizing corrective action.
3. **Correct:** Develop alternative corrective actions and implement the most appropriate action based on the seriousness of the potential failure and the project goals and constraints. Assess the success of the corrective actions taken.

23.2.1 Understand Potential Failure Modes and Their Effects

In the "understand" phase of the FMEA process, the team seeks answers to the following key questions:

1. What is the intended function of the component, subsystem, or system?
2. What are the possible failure modes? That is, how could the component, subsystem, or system conceivably fail to perform its intended function?
3. What would be the effect if the failure did occur?
4. What mechanisms or causes might produce these failure modes?
5. What current controls or counter-measures are provided to prevent the failure or to compensate for it?

The intended *function* is the task that the component, subsystem, or system must perform. Functions should be concisely described in a way that is easy for all users to understand. Functions are typically stated as actions such as: position, support, seal in, retain, lubricate, and so forth. *Failure* is any change in a product component, subsystem, or system that causes it to be unable to satisfactorily perform its intended function or to perform to design intent.

Failure modes are the ways in which the component, subsystem, or system could fail to perform its intended function. A component could fail by cracking, deforming, melting, leaking, interfering with other components, corroding, sticking, rubbing, short circuiting, open circuiting, vibrating, etc. Chrysler (1986) suggests that failure modes can be thought of as functional requirements expressed negatively. One way to identify failure modes is to ask what could happen to cause loss of function. For example, the failure modes of an electrical switch could be listed as:

- Fails to switch on
- Fails to switch off
- Slow or restricted action
- Inadvertent operation
- Short circuit to ground

Effect(s) of failure are the effects that the failure mode has on the function, as perceived by the customer. Effects should be described in terms of what the customer might notice or experience as a result of the failure mode. All customers should be considered including internal customers, external customers, and the ultimate end user. If the loss of function could impact safety or code compliance, this is a very serious effect that should also be clearly stated.

Mechanism(s) and cause(s) of failure are the possible mechanisms and/or causes of each failure mode. Identifying all possible mechanisms and causes is an important part of the FMEA because it points the way towards preventative or corrective action. An example of a cause of failure for the mode "fails to switch on" of an electrical switch might be "oxidation or corrosion of the contact surfaces resulting in poor electrical conductance." Other examples of design-related causes of failure include: insufficient wall thickness, incorrect tolerance specification, overloading of components, improper use of product, insufficient lubrication, contamination, vibration, shock loads, inadequate maintenance, and so forth.

Current *controls* or *countermeasures* are controls or preventative measures that are currently used to guard against a particular failure mode. They include all prevention steps, design verification and validation (DV) activities, or other activities that are intended to insure design adequacy with respect to the failure mode under consideration. Examples include design guidelines, specific rig or lab tests, drawings checked by a checker before release, design reviews, fail-safe features (e.g., pressure relief valve), materials engineering sign off, specification standards (e.g., weld specification), mathematical studies, computer simulations, and so forth. There are three types of controls that are typically employed:

Type 1: prevent the cause(s) of failure from occurring.

Type 2: detect the mechanism(s) or cause(s) of a failure mode and take corrective action.

Type 3: detect the failure mode.

Note that these types of control are listed in order of preference. Type 1 controls should be used whenever possible.

23.2.2 Assess Risk

The purpose of the risk assessment is to measure or establish the level of concern or seriousness associated a given failure mode. The assessment is implemented by decomposing the over all risk into smaller, easier-to-assess components, where each component is a risk factor. The procedure is as follows:

1. Determine the risk factors involved.
2. Decide on a point scale for assessing the risk associated with each factor. In all cases, a low number corresponds to low risk and a high number corresponds to high risk.
3. Calculate the overall risk, R, as the product of the point scores assigned to each factor, r_i, i.e.,

$$R = \prod_{i=1}^{n} r_i \qquad (23.1)$$

   where n is the total number of risk factors.
4. Decide on the appropriate interpretation or action for the resultant overall risk value.

Three risk factors are commonly considered in the FMEA. These are:

Severity (S): Severity is an assessment of the seriousness of the effect produced by the potential failure mode if it occurs. Severity applies to the effect only. As a general rule, the only way the severity rating can be reduced is by making an appropriate design change.

Occurrence (O): Occurrence is an estimation of the likelihood that a specific mechanism or cause will occur. Eliminating or controlling one or more of the mechanisms or causes of the failure mode through a design or process change is typically the only way the occurrence rating can be reduced.

Detection (D): Detection is an assessment of the ability of the current Type 2 control to detect a potential mechanism or cause, or the ability of a Type 3 control to detect the subsequent failure mode, before the component, subsystem, or system reaches the customer. In general, the detection rating can only be reduced by improving the planned control.

The team should agree on an evaluation criteria and point scale to be used for rating each risk factor. To ensure consistency, it is important that the same evaluation criteria and point scale be used for the duration of the project. A representative evaluation criterion based on a 1 to 5 point scale is shown in Table 23.1.

Using the agreed upon rating criteria, the overall risk is computed using Eq. (23.1) as

$$RPN = (S) \times (O) \times (D) \tag{23.2}$$

where *RPN* stands for "risk priority number." Having computed the *RPN* value, the team must agree on how it should be used or interpreted. Typically, the *RPN* value is used to rank order the design concerns. If the 5 point scale suggested in Table 23.1 is used, the *RPN* values will be between "1" and "125". Obviously an *RPN* value of "125" would be of great concern while a "1" could probably be completely ignored. The team must decide how RPN values that lie between these extremes are handled. One possible approach is illustrated in Table 23.2. As a general rule, failure modes that have a high severity rating should be given special attention regardless of the resultant *RPN* value.

23.2.3 Identify and Implement Corrective Actions

The intent of corrective action is to reduce any one or all of the risk factors. Typically, possible corrective actions are identified by "brainstorming" or by investigating the actions that were taken for a previous product or similar situation. Note that a design change can reduce any or all three of the risk factors and that a design change is the only way that the severity rating and occurrence rating can be reduced. The detection rating, on the other hand, can be reduced by making a design change, or in some cases, by increasing preventative measures or by additional validation and/or verification testing.

Table 23.1 Example design FMEA rating system
(MP = mechanical or electromechanical product; EP = electronic product)

Rating	Severity (S)	Occurrence (O)	Detection (D)
1	No effect to minor; Defect may be noticed by customer	Rare (<1 in 10^4 MP) (<1 in 10^6 EP)	Will almost certainly be detected
2	Customer is inconvenienced	Infrequent (2 to 10 in 10^4 MP) (2 to 10 in 10^6 EP)	Reasonably detectable by current controls
3	Item is operable, but at a reduced performance level; Customer is dissatisfied	Frequent (11 to 25 in 10^4 MP) (11 to 25 in 10^6 EP)	Detectable before reaching the customer
4	Item is inoperable; Loss of function	Very Frequent to High (26 to 50 in 10^4 MP) (26 to 50 in 10^6 EP)	Detectable only by the customer &/or during service
5	Safety-related catastrophic failure; Regulatory noncompliance involved	High to Very High (>50 in 10^4 MP) (>50 in 10^6 EP)	Undetectable until catastrophe occurs

Table 23.2 Example *RPN* interpretation criterion

RPN Value	Interpretation/Action
$1 \leq RPN \leq 17$	**Minor Risk:** little or no action is required.
$18 \leq RPN \leq 63$	**Moderate Risk:** This requires selective product validation and evaluation of the design and/or redesign to reduce the Risk Priority Number.
$64 \leq RPN \leq 125$	**Major Risk:** This should be given high priority. Extensive design revisions and other measures should be taken to reduce the Risk Priority Number.
Note: 1) If O or D $\geq$ 4, corrective action must reduce this to 3 or less. 2) If S = 5, both O and D must be 2 or less.	

Failure Mode and Effects Analysis											Sheet No. ___ of ___

System
Subsystem
Component

Description:

Comments:

1	2	3	4	5	6	7	8	9	10	11	Action Results
Function	Failure Mode	Effect(s) of Failure	Mechanism(s)/ Cause(s) of Failure	Current Controls	S	O	D	R P N	Recommended Corrective Action(s)	Actions Taken	S O D R P N

Figure 23.1 FMEA worksheet.

23.3 FMEA PROCEDURE

The FMEA is implemented by systematically completing a FMEA worksheet. Each firm or industry will typically have its own customized version of this worksheet. A representative worksheet is illustrated in Fig. 23.1. To perform the analysis, preliminary information is first recorded. This would include a description of the component, subsystem, or system that is being analyzed as well as other relevant information such as the reason for the analysis, the team members, and the date the analysis was performed. Note that many customized FMEA worksheets will provide separate spaces for much of this information. The analysis is performed by completing each column of the worksheet. A column-by-column guideline for performing the analysis is as follows:

1. **Function:** Enter, as concisely as possible, the function of the item being analyzed. Include information regarding the environment in which the item operates (e.g., temperature, pressure, humidity ranges). If the item has different functions with different potential failure modes, list all the functions separately.

2. **Failure Modes:** List each potential failure mode for the particular item and item function. Assume that the failure could occur, but may not necessarily occur. Potential failure modes can be identified by reviewing past failures of similar designs or situations, concerns generated in design reviews or otherwise expressed, and team "brainstorming."

3. **Effect(s) of Failure:** Describe the effects of the failure in terms of what the customer might notice or experience. Remember to consider all customers including internal customers as well as the ultimate end user. State clearly if the effect could impact safety or regulatory compliance.

4. **Mechanism(s)/Cause(s) of Failure:** List every conceivable failure mechanism or cause for each failure mode. Each mechanism or cause should be stated as concisely and completely as possible so that the most appropriate remedy can be identified.

5. **Current Controls:** List the prevention steps, design verification or validation activities, or other activities that are currently used to guard against each mechanism or cause. State these as concisely and completely as possible and clearly indicate the type of design control (Type 1, 2, or 3) that applies.

6. **Severity (S):** Assess the seriousness of the effect using the agreed upon point scale and assessment criteria (e.g., Table 23.1). Remember that the severity assessment applies to the effect only.

7. **Occurrence (O):** Assess the likelihood that each potential mechanism or cause will occur. Use the agreed upon point scale and assessment criteria (e.g., Table 23.1).

8. **Detection (D):** Assess the ability of the current controls to detect each potential mechanism or cause. Use the agreed upon point scale and assessment criteria (e.g., Table 23.1).

9. **Risk Priority Number (RPN):** Determine the RPN by computing the product of the severity rating (S), occurrence rating (O), and detection rating (D).

10. **Recommended Corrective Actions:** Develop alternative corrective actions based on the agreed upon interpretation of the Risk Priority Number (RPN) value (e.g., Table 23.2). Identify alternatives by reviewing past practices and by team "brainstorming" and select the most appropriate action based on project goals and impact on product cost, product performance, development cost, and development time. If no actions are recommended for a specific mechanism or cause, indicate this by entering "NONE" in this column. Also state the name of the individual responsible for implementing the corrective action and the target completion date.

11. **Action Results:** After an action has been implemented, enter a brief description of the actual action taken and calculate the resultant RPN.

Failure Mode and Effects Analysis										Sheet No. ___ of ___				

System	Description:	Comments:
X Subsystem	Motor Power Interrupt	Device senses motor stall or broken blade and interrupts power to
Component	Part # xxx	band saw motor.

1	2	3	4	5	6	7	8	9	10	11	Action Results			
Function	Failure Mode	Effect(s) of Failure	Mechanism(s)/ Cause(s) of Failure	Current Controls	S	O	D	R P N	Recommended Corrective Action(s)	Actions Taken	S	O	D	R P N
Interrupts motor power • saw dust • 30° C	Fails to actuate when triggered	Motor continues to operate under unsafe conditions • Endangers operator • UL reqm't	Binding linkage due to tolerance stack-up	Tolerance Stack Analysis Functional Test	5	2	3	30	Statistical tolerance stack analysis + natural tolerance (Type 2) Work with supplier to improve process (Type 2)	Change tolerance Specification Variability reduced by working with supplier	5	1	3	15
			Broken tension spring	Material Sign Off Prototype Testing	5	3	4	60	Redesign linkage so broken spring actuates device (Type 1) Test batch of springs for static strength and fatigue (Type 3) Use better material (Type 2)	Linkage redesigned to fail-safe Springs tested Better spring material specified	2	1	2	4
			Trip lever installed incorrectly	None	5	3	3	45	Redesign to make incorrect installation impossible (Type 1)	Lever mounting redesigned to prevent incorrect installation	1	1	1	1

Figure 23.2 Partially completed FMEA worksheet for a band saw motor emergency power interruption device.

23.4 SIMPLE EXAMPLE

To illustrate a typical FMEA worksheet, consider the partially completed analysis for an emergency band saw power interruption device shown in Fig. 23.2. The purpose of this device is to interrupt power to the band saw motor if it stalls or if the blade breaks. This sample worksheet illustrates several aspects of the FMEA process. To perform the FMEA, the team has to decide on a point scale and rating criterion which, for this example is the same as that given in Tables 23.1 and 23.2. Three mechanisms or causes of the "fails to actuate" failure mode are listed. Note that a risk assessment is performed for each failure mechanism or cause. Note also that only the risk factor effected by the action taken is changed. For example, by changing the tolerance specification and working with the supplier to reduce component to component variability, the probability of occurrence of linkage binding is reduced, but both the severity and probability of detection are unchanged. Similarly, by using a better spring material, the probability of the occurrence of a broken spring is reduced from 3 to 2. In addition, by changing the design so that a broken spring causes the device to actuate, the severity rating is reduced from "5" to "2" since the effect of a broken spring now results in customer inconvenience rather than danger.

Finally, by performing the FMEA, the team discovers that the "trigger lever" can be installed incorrectly. This illustrates one of the primary goals of FMEA, which is to identify and correct potential problems before they reach the customer.

23.5 KEY TAKEAWAYS

- Although experienced designers instinctively evaluate their designs with respect to possible failures, the evaluation process is, at best, haphazard and uncertain.
- FMEA brings discipline and structure to the process of evaluating a design for potential failures by providing a systematic and unambiguous analysis procedure.
- Because FMEA can be tedious and time consuming, its use requires company-wide support and nurturing.
- FMEA reduces total cost by helping the firm to avoid the warrantee costs, service calls, customer dissatisfaction, and damaged reputation that invariably results when quality and reliability problems reach the customer.

24

Design Review Checklist

24.1 INTRODUCTION

In this chapter, we present a comprehensive design review checklist that can be used by the design team to systematically question all aspects of the product design. This checklist is predicated on the assumption that quality is the summation of many important details, each carefully considered and decided upon during the design process. By systematically and conscientiously questioning the design as part of a formal design review process, the design team achieves "quality by design" by avoiding potential problems that might otherwise go unnoticed until they surface later as poor quality.

24.2 CUSTOMER SATISFACTION ASSESSMENT

___ **Are customer requirements, wishes, wants, and delights addressed by this design?**
  1. List the customers or users of the product or device.
  2. List the major "wants" for each user group (e.g., customer needs matrix, Fig. 8.5).
  3. Indicate how these wants are met by this design, or give rationale for why they are not.

___ **Does this design have aesthetic appeal?**
  1. List all steps taken to insure aesthetic appeal (e.g., focus groups).
  2. List all issues and/or concerns.
  3. Indicate how each issue has been resolved. If an issue is unresolved, provide status, rationale, or contingency plan.

___ **Are customer desires adequately translated into engineering specifications?**
1. List customer desires.
2. List the corresponding engineering specifications (e.g., house of quality, Fig. 8.6).
3. Indicate how each engineering specification fulfills customer desires.

___ **Have customer-related problems with existing products been addressed with this new design?**
1. List major customer problems with existing products.
2. Indicate how this design eliminates and/or addresses these problems.

___ **Does this design compare favorably with competitive products?**
1. Compare this design with "best in class" competitive product(s).
2. Clearly demonstrate how this design is better (e.g., concept selection matrix, Fig. 10.6).
3. Provide rationale for aspects that are not clearly better.

___ **Is this design easy for customers to use; is it "user-friendly"?**
1. List all "ease of use" issues associated with the design.
2. Indicate how each issue has been resolved. If an issue is unresolved, provide status, rationale, or contingency plan.

___ **Is the packaging appealing, and does it communicate important information?**
1. List all "packaging" issues associated with the design.
2. Indicate how each issue has been resolved. If an issue is unresolved, provide status, rationale, or contingency plan.

___ **Does this design meet all performance requirements and/or "must have" requirements?**
1. List all "performance" or "must have" issues associated with the design.
2. Indicate how each issue has been resolved. If an issue is unresolved, provide status, rationale, or contingency plan.

___ **Does this design meet its weight and/or size requirements?**
1. List all weight and size issues associated with the design.
2. Indicate how each issue has been resolved. If an issue is unresolved, provide status, rationale, or contingency plan.

24.3 BUSINESS ASSESSMENT

_____ **Have environmental, recycling, and other ecological considerations been given adequate attention?**
1. List all relevant environmental, recycling, and other ecological issues associated with the design.
2. Indicate how each issue has been resolved. If an issue is unresolved, provide status, rationale, or contingency plan.

_____ **Have safety issues and other legal and regulatory issues been satisfactorily addressed with this design?**
1. List all relevant safety, legal, and regulatory issues associated with the design.
2. Indicate how each issue has been resolved. If an issue is unresolved, provide status, rationale, or contingency plan.

_____ **Is this design easy to service and repair as determined by service personnel?**
1. List all "service and repair" issues associated with the design.
2. Indicate how each issue has been resolved. If an issue is unresolved, provide status, rationale, or contingency plan.

_____ **Does this design meet project objectives?**
1. List all product cost, product performance, development cost, and development time issues associated with the design.
2. Indicate how each issue has been resolved. If an issue is unresolved, provide status, rationale, or contingency plan.

_____ **Have manuals and other documentation been prepared and verified?**
1. List all documentation required.
2. Indicate status of each and plan for completing/correcting.

_____ **Has a contingency plan been considered to address unforeseen problems with new technologies?**
1. List all new technologies used in the design.
2. List the possible problems associated with each new technology?
3. List the contingency plans for each problem.

_____ **Does this design fit the product family architecture?**
1. List all nonconformances with the product architecture.

2. For each nonconformance, provide rationale and a plan for adapting the product into other products and applications.

___ **Have standard components selected from short lists of rationalized alternatives been used wherever possible?**
1. Identify all nonstandard or "odd-ball" components used in the design.
2. Provide rationale for each exception.

___ **Is this design compatible with foreseeable change?**
1. Evaluate the design with respect to the following questions:
   a) How might the product change over time? How might customer needs or functional requirements change? What applicable new technologies are likely to become available? How would these changes or developments effect the design?
   b) How might the process or production technology change over time. What effect would these changes have on the design?
   c) What product or model variations are planned? How does the product/process concept accommodate these variations? What new variations could be introduced in the future? How would these changes impact the product design and process plan?
2. Indicate how the design accommodates possibilities with respect to each question.

24.4 ROBUSTNESS ASSESSMENT

___ **Is this design concept free of undesirable interactions?**
1. Analyze the design for possible undesirable couplings and interactions and list problems that may result.
2. Provide a redesign and/or contingency plan for each problem.

___ **Are reliable, time-tested designs and components used?**
1. Identify all new or unproven designs and components used.
2. List possible problems with each new design or component.
3. Provide a contingency plan for each problem.

___ **Is this design robust with respect to hard-to-control factors?**
1. List the hard-to-control environmental factors (e.g., temperature, humidity), use factors (e.g., wear, noise), and manufacturing factors (e.g., dimensional variation) for the product, subsystems, and components.

2. Show how the controllable factor values (design parameters) and tolerances have been selected to maximize robustness (e.g., Taguchi method, probabilistic analysis, prototype testing, etc.).

_____ **Have robust materials been selected?**
Do the following for each material specification used in the design:
1. List key material requirements including range of acceptable values. Requirements should include desired properties, availability, supplier quality, cost, etc.
2. List candidate materials considered.
3. Provide rational for each material selected.

24.5 MANUFACTURABILITY ASSESSMENT

_____ **Have manufacturability-related problems with existing products been addressed with this new design?**
1. List major manufacturability problems with existing products.
2. Indicate how this design eliminates and/or addresses these problems.

_____ **Has manufacturing indicated that they have the capability to manufacture this design?**
1. List all "in-house" manufacturing processes required, indicate whether the process is new to the company, and determine existing manufacturing capacity.
2. Provide rational for selecting each process that is new to the company.
3. Provide a plan for all processes that have insufficient existing capacity.

_____ **Have all suppliers indicated that they have the ability to manufacture this design?**
Do the following for each major purchased component:
1. List all suppliers. Assess each supplier's capability to meet quality, quantity, and schedule requirements. Provide a brief explanation for each assessment including discussion of open issues.
2. Indicate what steps are being taken to insure each supplier's performance.
3. Provide contingency plans for each problem or issue identified.

___ **Has adequate consideration been given to using existing equipment, fixtures, and tooling common with other products?**
Do the following for the product, major subsystems, and components:
1. List all equipment, fixtures, and tooling required and indicate if it is special purpose or common with other products.
2. Provide a rational for each special purpose piece of equipment, fixture, or tool.

___ **Is the design easy to manufacture and assemble using the fewest parts and fasteners possible?**
Do the following for the product, major subsystems, and assembled components:
1. List the number of designed parts, the total number of parts, and indicate which parts are candidates for elimination (CFE) based on the questions of motion, material, and assembly/service. Provide a brief rational for each CFE.
2. List, describe briefly, and provide a rationale for all violations of the design for manufacture and assembly guidelines.
3. If available, provide assembly efficiency and other results of DFA analysis.

___ **Are components overconstrained in this design?**
1. List all subassemblies and components that are overconstrained by their method of assembly or attachment.
2. For each instance of overconstraint, list the possible problems that may result.
3. For each potential problem, list current controls and proposed corrective actions.

___ **Does this design have critical tolerance stack-ups?**
Do the following for the final product and each subassembly:
1. List all critical tolerance stack-ups.
2. For each tolerance stack-up, list the possible problems that may result.
3. For each potential problem, list current controls and proposed corrective actions.

___ **Is this design easy to test?**
1. List all tests to be performed as part of the production process.
2. List issues and problems associated with each test.
3. Indicate how each issue has been resolved. If an issue is unresolved, provide status, rationale, or contingency plan.

24.6 RELIABILITY ASSESSMENT

____   **Have all "performance critical" and "safety critical" aspects and characteristics of the design been properly identified?**
Do the following for the product, subsystems, and components:
1. List all performance and safety critical characteristics; briefly describe issues and concerns.
2. Indicate all steps and measures that have been taken to resolve or mitigate potential problems, including analysis, testing, etc.

____   **Has the combined effect of force, reactive environment, time, and temperature been considered?**
Answer the following questions in detail for the product, subsystems, and components.
1. Are the materials selected compatible with the expected corrosive environment?
2. Are surface treatments acceptable?
3. If two or more metals are connected by an electrolyte, are the metals as close together as possible in the galvanic series? If not, has an appropriate insulator been used to break the galvanic circuit? Has a sacrificial surface been considered? Can surfaces drain easily?
4. Are crevices avoided where possible? Are unavoidable crevices properly sealed? Can surfaces that collect corrosives be readily cleaned?
5. Has stress-corrosion cracking and corrosion fatigue been properly considered by using more corrosion resistant materials, reducing corrosive action (e.g., protective coatings, inhibitors, cathodic protection, etc.), reducing tensile stress (e.g., reducing interference fits, etc.) and introducing beneficial compressive stress (e.g., shot-peening, etc.).
   a) List all instances where stress-corrosion or corrosion fatigue is a concern.
   b) List current controls and proposed corrective actions for each.
6. Have adhesive wear, abrasive wear, and corrosion film wear been properly considered? Is scuffing, scoring, gulling, or fretting a concern?
   a) List all instances where wear or wear failure is a concern.
   b) List current controls and proposed corrective actions for each.

____   **Have critical-load-carrying members been designed such that the force flow is smooth and gradual in regions of high stress?**
1. List all aspects or features of the design where stress concentration, residual stress, or other "weak links" is a concern.
2. List all instances where consideration of smooth force flow has been a consideration in determining the shape of a designed component. Examples include optimizing the part shape using FEA, placing stress raisers in regions of low nominal stress, using large radii at fillets, grooves, holes, etc., and general smoothing of contours around necessary changes in section.
3. Provide a rational for each instance involving stress raisers or other undesirable effects that have not been avoided or designed for smooth force flow.

____   **Have proven mechanical design practices been used wherever possible in the design of fabricated or built-up structures?**
1. Evaluate the design with respect to the following practices:
    a) To avoid localized buckling or structural failure, is the strength of each redundant load-carrying path approximately proportional to its stiffness?
    b) Has the stiffest redundant load-carrying path been made significantly stiffer than all other paths to minimize the effect of manufacturing process induced residual stress?
    c) Is the stiffest redundant load-carrying path also the most precise and dimensionally accurate to ensure high quality of the final assembly?
    d) To avoid stress concentration due to stiffness mismatch, have related load-carrying components been designed such that under load they deform in the same sense, and if possible, by the same amount?
2. List all instances where a practice could have been applied but wasn't. Explain the rational for not using the proven practice for each instance. List current controls and proposed corrective actions for each.

____   **Has adequate protection against shipping and storage damage been provided?**
1. List all ways damage can occur during shipping and storage.
2. Indicate how each possibility is guarded against.
3. Provide rational for all possibilities that are not guarded against. List current controls and proposed corrective actions for each.

___ **Is this design easy to maintain?**
Provide detail answers to the following questions:
1. Have on-line maintenance possibilities been maximized?
2. Is there a risk of damage when performing routine maintenance and how is this guarded against?
3. Can maintenance points be accessed without disturbing other components and attachments such as piping or electrical cabling?
4. Has interchangeability of parts been maximized?
5. Is there a risk of damage through installation error and how has this been guarded against?
6. Has the design been optimized for quick maintenance (e.g., few fasteners, fast-acting fasteners, easy assembly, simple adjustments, quick-change replaceable modules, etc.)?
7. Are special skills or tools required for maintenance, and if so, why haven't they been avoided by design?
8. Is there an active plan to minimize MTTR?

___ **Are wear, deterioration, and potential failures "detectable"?**
Provide detail answers to the following questions:
1. Can wear progress of parts be monitored? How or why not?
2. Is there advance warning of failure? How or why not?
3. Can deterioration be detected and predicted? How or why not?
4. Is accessibility for on-line inspection provided? How or why not?
5. Is there a risk of damage when performing on-line inspection and how is it guarded against?
6. Can important performance indicators be easily observed? How or why not?

24.7 KEY TAKEAWAYS

- The design review checklist provides discipline and objectivity for systematically evaluating the readiness of a product design for release to manufacture and for identifying potential trouble spots that require further consideration.
- The checklist provides a comprehensive guide that can be used by the team to prepare for a design review. It can also be used by the firm as a basis for conducting design reviews.
- The checklist should be modified and continuously improved by including additional considerations and deleting those that don't apply.
- Many of the chapters in this book can be used as sources for additional questions and considerations for inclusion in the checklist.

References

AlliedSignal, 1994, *A Guide to Designing and Manufacturing Environmentally Friendly Products*, AlliedSignal Inc., 101 Columbia Road, Morristown, NJ 07962.

Arora, J., 1989, *Introduction to Optimum Design*, John Wiley & Sons, New York.

Beall, G., 1985, *Solving the Plastic Product Design and Development Puzzle*, workshop notes presented by Glenn Beall Plastics Limited, 32981 North River Road, Libertyville, IL 60048.

Bendell, A., Disney, J., and Pridmore, W., 1989, *Taguchi Methods: Applications in World Industry*, IFS Publications, Bedford, UK and Springer-Verlag, Berlin.

Bloch, H. and Geitner, F., 1990, *An Introduction to Machinery Reliability Assessment*, Van Nostrand Reinhold, New York, NY, p.132.

Boothroyd, G., 1992, *Assembly Automation and Product Design*, Marcel Dekker, Inc., New York.

Boothroyd, G., Dewhurst, P., and Knight, W., 1994, *Product Design for Manufacture and Assembly*, Marcel Dekker, Inc., New York.

Bradyhouse, R., 1987, "The Rush for New Products Versus Quality Designs That Are Producible; Are These Objectives Compatible?" presented at the *SME Simultaneous Engineering Conference*, held June 1, 1987, Society of Manufacturing Engineers, Dearborn, MI.

Burns, M., 1993, *Automated Fabrication*, PTR Prentice Hall, Englewood Cliffs, NJ.

Camp, R., 1989, *Benchmarking*, ASCQ Quality Press, Milwaukee, WI.

Chironis, N. and Sclater, N., 1996, *Mechanisms and Mechanical Devices Sourcebook*, 2nd Ed., McGraw-Hill, New York.

Chow, W., 1977, "Snap-Fit Design," *Mechanical Engineering*, July 1977, pp. 35-41.

Chow, W., 1978, *Cost Reduction in Product Design*, Van Nostrand Rheinhold Company, New York.

Chrysler Motors, 1986, *Failure Mode and Effects Analysis Manual*, prepared by Design Feasibility & Reliability Assurance, Chrysler Motors, Detroit, MI.

Conley, J. and Stoll, H., 1995, "Prototyping and Tooling for Rapid Product Development," two-day seminar presented by Northwestern University, McCormick School of Engineering and Applied Science, May 15-16, 1995.

Cross, N., 1994, *Engineering Design Methods*, 2nd Ed., John Wiley & Sons, New York.

Dixon, J. and Poli, C., 1995, *Engineering Design and Design For Manufacturing*, Field Stone Publishers, Conway, MA.

Ettlie, J. and Stoll, H., 1990, *Managing the Design-Manufacturing Process*, McGraw-Hill, New York.

Fallon, C., 1980, *Value Analysis*, Prentice Hall, Englewood Cliffs, NJ.

Ford Motor Company, 1985, "Automation Friendly Design," two-day course presented by the Robotics and Automation Applications Consulting Center, Ford Motor Company, June 27-28, 1985.

Fowler, T., 1990, *Value Analysis in Design*, Van Nostrand Reinhold, New York.

Garvin, D., 1987, "Competing on the Eight Dimensions of Quality," Harvard Business Review, Nov/Dec 1987, p 101-109.

Gerber, B., 1990, "Benchmarking: Measuring Yourself Against the Best," Training, v27n11, Nov. 1990, pp. 36-42.

Gordon, W., 1961, *Synectics: the Development of Creative Capacity*, Harper and Row Pub., New York.

Groover, M., 1987, *Automation, Production Systems, and Computer-Integrated Manufacturing*, Prentice-Hall, New Jersey, p.40.

Groover, M., 1996, *Fundamentals of Modern Manufacturing*, Prentice-Hall, Englewood Cliffs, NJ.

Haugen, E., 1980, *Probabilistic Mechanical Design*, John Wiley and Sons, New York.

Hoult, D. and Meador, C., 1996, "Predicting Product Manufacturing Costs from Design Attributes: A Complexity Theory Approach," No. 960003, Society of Automotive Engineers.

Jay, F., 1971, "Joining Parts by Die-Casting," *Machine Design*, April 15, 1971, pp. 35-41.

Juvinal, R., 1983, *Fundamentals of Machine Component Design*, John Wiley & Sons, New York, pp. 40-42.

Kaplan, R. (ed.), 1990, *Measures of Manufacturing Excellence*, Harvard Business School Press, Boston, MA.

Kriegel, J., 1995, "Exact Constraint Design," Mechanical Engineering, V117, N5, May 1995, pp. 88-90.

Kuester, J. and Mize, J., 1973, *Optimization Techniques with Fortran*, McGraw-Hill, New York.

Lochner, R. and Matar, J., 1990, *Designing for Quality*, ASQC Quality Press, Milwaukee, WI.

McMahon, C. and Browne, J., 1993, *CADCAM: From Principles to Practice*, Addison-Wesley, Reading, MA.

Military Handbook 727, 1984, *Design Guidance for Producibility*, Department of Defense, Washington, DC.

Muccio, E., 1991, *Plastic Part Technology*, ASM International, Materials Park, OH 44073-0002.

Osborn, A., 1979, *Applied Imagination*, Charles Scribner's Sons, New York.

Ostrofsky, B., 1977, *Design, Planning, and Development Methodology*, Prentice-Hall, Englewood Cliffs, NJ.

Ostwald, P., 1988, *American Machinist Cost Estimator*, 4th ed. Penton Publishing, Cleveland, OH.

Pahl, G. and Beitz, W., 1988, *Engineering Design*, K. Wallace (ed.), Springer-Verlag, Berlin, pp. 199-203.

Papalambros, P. and Wilde, D., 1988, *Principles of Optimal Design*, Cambridge University Press, Cambridge, England.

Peace, G., 1993, *Taguchi Methods*, Addison-Wesley, Reading, MA..

Pradke, M., 1989, *Quality Engineering Using Robust Design*, Prentice-Hall, Herts, England.

Pugh, S., 1991, *Total Design: Integrating Methods for Successful Product Engineering*, Addison-Wesley, Reading, MA.

Raudsepp, E., 1983, "Stimulating Creative Thinking," Machine Design, June 9, 1983, pp. 75-78.

Redford, A. and Chal, J., 1994, *Design for Assembly Principles and Practice*, McGraw-Hill, New York.

Ross, P., 1988, *Taguchi Techniques for Quality Engineering*, McGraw-Hill, New York.

Schey, J., 1987, *Introduction to Manufacturing Processes*, McGraw-Hill, New York.

Shah, J. and Mantyla, M., 1995, *Parametric and Feature-Based CAD/CAM*, John Wiley & Sons, Inc., New York.

Shigley, J. and Mischke, C., 1989, *Mechanical Engineering Design*, McGraw-Hill, New York.

Suh, N., Bell, A., and Gossard, D., 1978, "On an Axiomatic Approach to Manufacturing and Manufacturing Systems," Journal for Engineering in Industry, vol. 100, no. 2, pp 127-130.

Suh, N., 1990, *The Principles of Design*, Oxford University Press, Oxford, England.

Taguchi, G. and Yuin, W., 1979, *Introduction to Off-Line Quality Control*, Central Japan Quality Control Association, Nagaya, Japan.

Tipping, W., 1969, *An Introduction to Mechanical Assembly*, Business Books Ltd., London, England.

Ullman, D., 1997, *The Mechanical Design Process*, 2nd Ed., McGraw-Hill, New York.

Ulrich, K. and Eppinger, S., 1995, *Product Design and Development*, McGraw-Hill, New York.

VDI Guideline 2221, 1987, "Systematic Approach to the Design of Technical Systems and Products," VDI Society for the Product Development, Design, and Marketing, VDI-Verlag GmbH, D-4000 Dusseldorf.

Watson, M., 1996, "A Standardization Analysis Process Applied to Steel Coils in the Automotive Industry," Doctoral Thesis, Northwestern University, Evanston, IL.

Wilde, D., 1978, *Globally Optimal Design*, John Wiley & Sons, New York.

Wilson, D., 1980, "An Exploratory Study of Complexity in Axiomatic Design," Doctoral Thesis, Massachusetts Institute of Technology, Cambridge, MA.

Yasuhara, M and Suh, N., 1980, "A Quantitative Analysis of Design Based on Axiomatic Approach," in *Computer Applications in Manufacturing Systems*, ASME Prod. Engr. Div. Publ., PED Vol. 2.

Zwicky, F., 1969, *Discovery, Invention and Research through the Morphological Approach*, Macmillan, New York.

Appendix

Table A.1 Probability $\alpha = P(Z > Z_\alpha)$ for a normal (Gaussian) distribution*

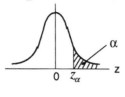

Z_α	0.00	0.01	0.02	0.03	0.04	0.05	0.06	0.07	0.08	0.09
0.0	0.5000	0.4960	0.4920	0.4880	0.4840	0.4801	0.4761	0.4721	0.4681	0.4641
0.1	0.4602	0.4562	0.4522	0.4483	0.4443	0.4404	0.4364	0.4325	0.4286	0.4247
0.2	0.4207	0.4168	0.4129	0.4090	0.4052	0.4013	0.3974	.03936	0.3897	0.3859
0.3	0.3821	0.3783	0.3745	0.3707	0.3669	0.3632	0.3594	0.3557	0.3520	0.3483
0.4	0.3446	0.3409	0.3372	0.3336	0.3300	0.3264	0.3238	0.3192	0.3156	0.3121
0.5	0.3085	0.3050	0.3015	0.2981	0.2946	0.2912	0.2877	0.2843	0.2810	0.2776
0.6	0.2743	0.2709	0.2676	0.2643	0.2611	0.2578	0.2546	0.2514	0.2483	0.2451
0.7	0.2420	0.2389	0.2358	0.2327	0.2296	0.2266	0.2236	0.2206	0.2177	0.2148
0.8	0.2119	0.2090	0.2061	0.2033	0.2005	0.1977	0.1949	0.1922	0.1894	0.1867
0.9	0.1841	0.1814	0.1788	0.1762	0.1736	0.1711	0.1685	0.1660	0.1635	0.1611
1.0	0.1587	0.1562	0.1539	0.1515	0.1492	0.1469	0.1446	0.1423	0.1401	0.1379
1.1	0.1357	0.1335	0.1314	0.1292	0.1271	0.1251	0.1230	0.1210	0.1190	0.1170
1.2	0.1151	0.1131	0.1112	0.1093	0.1075	0.1056	0.1038	0.1020	0.1003	0.0985
1.3	0.0968	0.0951	0.0934	0.0918	0.0901	0.0885	0.0869	0.0853	0.0838	0.0823
1.4	0.0808	0.0793	0.0778	0.0764	0.0749	0.0735	0.0721	0.0708	0.0694	0.0681
1.5	0.0668	0.0655	0.0643	0.0630	0.0618	0.0606	0.0594	0.0582	0.0571	0.0559
1.6	0.0548	0.0537	0.0526	0.0516	0.0505	0.0495	0.0485	0.0475	0.0465	0.0455
1.7	0.0446	0.0436	0.0427	0.0418	0.0409	0.0401	0.0392	0.0384	0.0375	0.0367
1.8	0.0359	0.0351	0.0344	0.0336	0.0329	0.0322	0.0314	0.0307	0.0301	0.0294
1.9	0.0287	0.0281	0.0274	0.0268	0.0262	0.0256	0.0250	0.0244	0.0239	0.0233
2.0	0.0228	0.0222	0.0217	0.0212	0.0207	0.0202	0.0197	0.0192	0.0188	0.0183
2.1	0.0179	0.0174	0.0170	0.0166	0.0162	0.0158	0.0154	0.0150	0.0146	0.0143
2.2	0.0139	0.0136	0.0132	0.0129	0.0125	0.0122	0.0119	0.0116	0.0113	0.0110
2.3	0.0107	0.0104	0.0102	.00990	.00964	.00939	.00914	.00889	.00866	.00842
2.4	.00820	.00798	.00776	.00755	.00734	.00714	.00695	.00676	.00657	.00639
2.5	.00621	.00604	.00587	.00570	.00554	.00539	.00523	.00508	.00494	.00480
2.6	.00466	.00453	.00440	.00427	.00415	.00402	.00391	.00379	.00368	.00357
2.7	.00347	.00336	.00326	.00317	.00307	.00298	.00289	.00280	.00272	.00264
2.8	.00256	.00248	.00240	.00233	.00226	.00219	.00212	.00205	.00199	.00193
2.9	.00187	.00181	.00175	.00169	.00164	.00159	.00154	.00149	.00144	.00139
Z_α	0.0	0.1	0.2	0.3	0.4	0.5	0.6	0.7	0.8	0.9
3	.00135	$.0^3968$	$.0^3687$	$.0^3483$	$.0^3337$	$.0^3233$	$.0^3159$	$.0^3108$	$.0^4723$	$.0^4481$
4	$.0^4317$	$.0^4207$	$.0^4133$	$.0^5854$	$.0^5541$	$.0^5340$	$.0^5211$	$.0^5130$	$.0^6793$	$.0^6479$
5	$.0^6287$	$.0^6170$	$.0^7996$	$.0^7579$	$.0^7333$	$.0^7190$	$.0^7107$	$.0^8599$	$.0^8332$	$.0^8182$
6	$.0^9987$	$.0^9530$	$.0^9282$	$.0^9149$	$.0^{10}777$	$.0^{10}402$	$.0^{10}206$	$.0^{10}104$	$.0^{11}523$	$.0^{11}260$

*Adapted from Shigley and Mischke, 1989.

376

Index